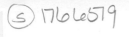

Fractals

Fractals

Endlessly Repeated
Geometrical Figures

HANS LAUWERIER

TRANSLATED BY

SOPHIA GILL-HOFFSTÄDT

PRINCETON UNIVERSITY PRESS

PRINCETON, NEW JERSEY

Lauwerier, H. A. (Hendrik Adolf)
[Fractals. English]
Fractals : endlessly repeated geometrical figures / Hans Lauwerier;
translated by Sophia Gill-Hoffstädt.
p. cm. — (Princeton science library)
Translation of: Fractals.
Includes bibliographical references and index.
ISBN 0-691-08551-X (cl.) 0-691-02445-6 (pbk.)
1. Fractals. I. Title. II. Series.
QA614.86.L3813 1991
514'.74—dc20   90-40842

CONTENTS

PREFACE

This book has been written for a wide audience, for anyone interested in fractals—whether graphic designer, computer fanatic, mathematician, natural philosopher, and so on, in random order. Everyone will find something to his liking. This isn't the kind of book to read or look at from cover to cover; it's a book to work with.

Readers who are good at working with microcomputers could start with the final chapter, try out one of the programs listed there, and only later consult the earlier chapters for whatever they need to understand the fractals made visible in this way.

Without a computer one can still have a good time, but in this case it is better to start with the first chapter.

Nearly all the pictures of fractals were made by the author on an Olivetti M240 with monochrome screen. The color pictures were made by Jaap Kaandorp using the more sophisticated equipment of the Centre for Mathematics and Computer Science (CWI) in Amsterdam.

The manuscript of this book was read in its early stages by Jan van de Craats and Klaas Lakeman. Their remarks led to many improvements. Klaas Lakeman also took care of the final editing.

Amsterdam, May 1987

Hans Lauwerier

ACKNOWLEDGMENTS

Nearly all the colored illustrations are reproduced by courtesy of the Centre for Mathematics and Computer Science, Amsterdam. They were produced by Jaap Kaandorp.

The fractal landscapes in Figures 6.6 and 6.7 are reproduced by kind permission of B. B. Mandelbrot. All other illustrations were made by the author.

The fractals in the book have been photographed on the computer screen or printed with a laser printer.

INTRODUCTION

A fractal is a geometrical figure in which an identical motif repeats itself on an ever diminishing scale. Whereas once they were just mathematical curiosities, fractals now receive a lot of attention. Their godfather is the Franco-American mathematician Benoit B. Mandelbrot. His book *The Fractal Geometry of Nature* contains a large number of illustrations, many of which were made with the help of a computer.

These days even simple microcomputers have good graphic possibilities, and many fractals described by Mandelbrot and others can be conjured up on the screen. In this book we will go into this in great detail. Anyone with a microcomputer such as the Olivetti M240 at his disposal, using the programming language TURBO BASIC (or POWER BASIC) and the operating system MS-DOS, can start working with the computer programs listed in Appendix B at once.

It is important to have a screen with a high resolution (640×400 pixels). Whether the screen is colored or not matters less. Of course fantastic results can be obtained with colors, as we'll also show from time to time. Our color illustrations, however, were produced with sophisticated apparatus. In their recent book *The Beauty of Fractals*, Peitgen and Richter demonstrate the spectacular results that can be obtained by working in color. This does not alter the fact that with more modest means we can follow them in nearly everything.

The great charm of all this lies not so much in looking at a single picture, but in watching the fractal actually coming into being on the screen. It's like making music yourself without possessing the technique of a professional musician. We do not, however, assume the reader or user of this book will restrict himself to computer work. By studying the various chapters he can also gain insight into the geometrical structure of fractals. To this end enough background information has been included to satisfy readers who do not have access to a computer.

The background information includes some mathematics, no more than necessary to get to the bottom of the mathematical structure of a fractal. Of course one can also enjoy fractals without mathematics;

one could see them as mysterious objects of "computer art." Only the prior knowledge and the studiousness of the reader determine to what extent he is able and willing to go into the underlying mathematics. A number of the more mathematical passages could be skipped at first reading without problems. Practically minded persons could start with Chapters 3 and 4, followed by Chapter 7. Anyone wanting to experiment with a computer should study Chapter 8 especially.

Fractals are characterized by a kind of built-in *self-similarity* in which a figure, the motif, keeps repeating itself on an ever-diminishing scale. A good example is a tree with a trunk that separates into two branches, which in turn separate into two smaller side branches, and so on. In our mind we can repeat this an infinite number of times. The final result is a tree fractal with an infinite number of branches; each individual branch, however small, can in its turn be regarded as a small trunk that carries an entire tree. This construction has a lot to do with the binary number system. In fact, that is the reason why we start this book with a chapter on number systems.

Anyone studying the computer programs will notice that occasionally a natural number, an index number, has been written down in a particular number system. This happens especially in programs such as MEANDER and STAR. The first chapter summarizes everything we need to know about number systems.

In the second chapter a number of basic mathematical concepts important for gaining a deeper insight into the nature of fractals come up for discussion. Anyone who prefers concrete examples to general theory can restrict himself to Cantor's point-set. Not only is this point-set the most ancient fractal (about a hundred years old), but it is now also found to be an essential ingredient of many modern fractals.

Many examples of fractals are included in Chapters 3 and 4. Recurring themes are the motif being repeated on an ever-reduced scale, the tree structure, and geometrical similarity. In the fourth chapter we will show how a repeated similarity transformation, a rotation and a rescaling combined, will lead to spiral forms. Sure enough, many fractals are full of spirals. As a matter of fact the same applies to the world around us, from seashell to spiral nebula.

If we define a fractal as a figure with a self-similarity that continues reducing indefinitely, then the spiral, or rather the logarithmic spiral, may be considered a kind of primeval fractal, from which many intricate fractals can be built up. The original definition given by Mandelbrot is more limited and not very manageable because it involves the concept of "fractal dimension." This is rather difficult to calculate. In Chapter 5 we will look into this mathematical concept.

All the same, this concept did lead to the name "fractal" (the Latin *fractus* means broken). In this book we prefer to emphasize geometrical self-similarity, which is why the emphasis of the fifth chapter lies both on the geometrical aspects of fractals and their symmetries and similarities. We regard the fact that a fractal has a fractal dimension as a coincidence. One thing and another leads to an extended class of fractals built up from separate points, so-called dust clouds. These are built up according to the principle of a binary or *n*-ary tree and subjected to an infinite sequence of similarity transformations.

Chapter 5 also includes a discussion of the backtrack method, a technique for efficiently implementing tree structures in computer programs that requires little memory space.

The concept "fractal" has already proved its use in many applied fields. There one often feels the need to extend the concept of similarity to some degree by introducing small changes into the series of similarity transformations, so-called disturbances. If we introduce chance disturbances into a mathematically regular tree fractal the result may look like a real tree, coral, or sponge.

In Chapter 6 we will show how chance can be incorporated in mathematical model-building. The computer is an ideal device for forming such so-called stochastic fractals. In the film industry this method is used to design imaginary landscapes in science-fiction films. We merely have to think of the self-similar structure of the surface of the moon with its craters of various sizes. In this chapter we look into the way the computer imitates chance, the concept of deterministic chaos. We will explain this concept using a simple mathematical model. This model describes, among other things, the restricted population growth of successive generations of insects. Precisely because it is so simple, mathematicians and physicists are extremely interested in it. Recently it led the physicist Mitchell Feigenbaum to the discovery of a natural law, a kind of universal similarity, with a scaling constant that is now called Feigenbaum's number.

Chapter 7 takes us right into the area of current research on fractals, associated with the names Poincaré, Julia, and Mandelbrot. The key phrase here is *dynamical system*. A dynamical system is a model for the motion of matter in a field of force. Here we are thinking not only of planets and comets in our solar system, but also of the interaction of elementary particles in an accelerator. A dynamical system reveals itself as a repeated transformation of a plane figure and leads to self-similar geometrical patterns—fractals.

The models we describe are very simple and lend themselves extremely well to computer experiments. It requires little effort to carry out these experiments on a simple computer, and one is rewarded by beautiful pictures of fractals. If more powerful computer apparatus is available, fascinating colored fractal patterns can be formed in just the same way, as items for an exhibition of "computer art." In places this chapter goes into the theoretical background fairly deeply to make it possible for readers with some mathematical knowledge not only to watch or make fractals but also to understand them.

Anyone who would rather skip or postpone the theory can go straight to Chapter 8, which has an entirely practical orientation. Designing fractals on the screen of a microcomputer yourself is the best way to get to know fractals. In this chapter we concentrate on a few very simple programs, on which many variations are possible. These are programs of the type MIRA and CLOUD for the dynamical systems we treat in Chapter 7, programs of the type DUST for the fractals in Chapter 5, and finally the programs of the type STAR and MEANDER for the fractals discussed in Chapters 3 and 4.

All programs can be used with TURBO BASIC without modifications; all that is needed is an appropriate choice of screen number. For those who want to use a different programming language, the specific points of the graphical interface of TURBO BASIC are summarized in the beginning of Chapter 8.

In this book we make only modest use of mathematical techniques. In principle everything is built on efficient use of coordinate geometry. We assume that the reader has some familiarity with trigonometrical functions, the exponential function, and logarithms. On the other hand, we do not assume that the reader knows anything about complex numbers. That is a pity, because with complex numbers many things could be expressed more compactly and elegantly, especially the fractals of Julia and Mandelbrot. Appendix A includes material suitable for the reader already familiar with complex numbers.

CHAPTER 1

Counting and Number Systems

A *fractal* is a geometrical figure that consists of an identical motif repeating itself on an ever-reduced scale. A good example of this is the H-fractal (Figure 1.1). Here the capital letter H is the repeating motif. The H-fractal is built up step by step out of a horizontal line-segment (located in the middle of Figure 1.1) taken to be of unit length. At the first step two shorter line-segments are placed perpendicularly at the ends of the original one. In Figure 1.1 a reduction factor of $1/\sqrt{2}$ has been chosen. At the second step, shorter horizontal line-segments are fastened on to the four endpoints in the same way. The same reduction factor makes the lengths of these half a unit. At the third step, eight vertical line-segments are added. We could continue like this for a long time.

Figure 1.1 stops with the tenth step. At that step 1 024 line-segments are added, each of length $1/32$. For practical reasons we can't carry on much longer. Nothing stops us, however, from imagining this process continuing indefinitely. The figure that emerges is an ideal mathematical fractal, in which every part—however small—is representative of the whole. To make this, an infinite number of steps are necessary. Figure 1.1 is a good approximation of such an ideal fractal. Approximations like this are usually called fractals too.

Tree Structure

One could interpret the H-fractal in Figure 1.1 as the plan of a town, a typical residential town, not suitable for through traffic. The moment the road is blocked somewhere the town falls apart into two separate parts. The mathematical term for this is "singly connected." The H-fractal is called a dendrite, after the Greek *dendron*, tree. This name is very apt because the structure of this fractal is indeed that of a tree. A trunk separates into two side branches, each of which acts as a trunk for the following two smaller side branches, and so on.

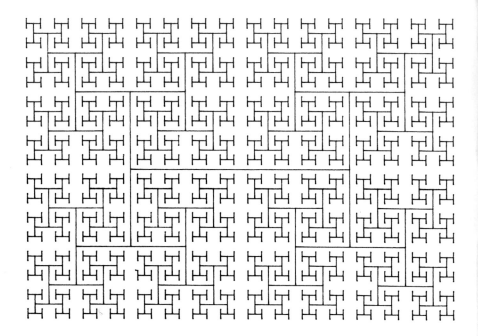

Figure 1.1 H-fractal

Figure 1.2 is an even simpler geometrical representation of the tree structure. Here there are, conceptually at least, infinitely many levels. At every level the vertical branches split in two, with a reduction factor of 1/2. The vertical branches double in number at every level, while their individual lengths are simultaneously halved. Each horizontal line is twice the length of the vertical branch below it. If we give the branch at the lowest level (the trunk) length 1, then vertically length 1 is added every time:

$$1 \times 1 + 2 \times \frac{1}{2} + 4 \times \frac{1}{4} + 8 \times \frac{1}{8} + 16 \times \frac{1}{16} + \cdots$$

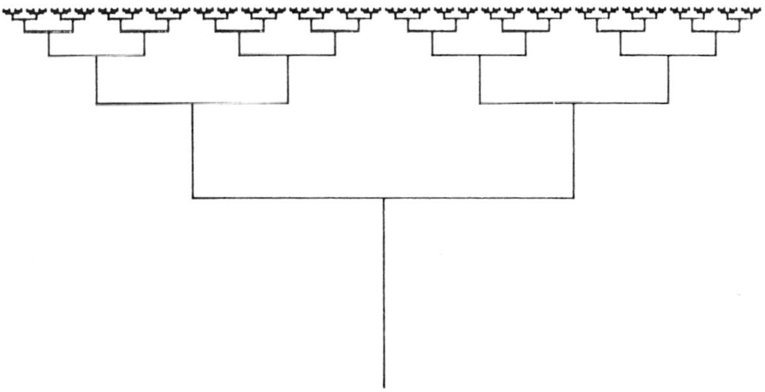

Figure 1.2 Binary tree

What strikes us especially in Figure 1.2 is the *self-similarity*. Each vertical branch, however small, can in its turn be considered the trunk of a complete tree, a scaled-down copy of the entire figure.

Another fact that strikes us is that the whole figure is confined within a right-angled triangle: half a square. As the branches get smaller, they come closer to the top edge. The picture shows, however, that they'll always be one branch-length away from it. We are dealing here with a *limit process* comparable to the summation of an infinite series:

$$1 + \frac{1}{2} + \frac{1}{4} + \frac{1}{8} + \frac{1}{16} + \frac{1}{32} + \cdots = 2.$$

Of course we can't carry on adding indefinitely. We always have to stop somewhere. We can nevertheless get arbitrarily close to 2, the limit value. After 5, 10, or 20 terms, for example, we find the following:

n	sum	difference from 2 (limit)
5	1.937 5	0.062 5
10	1.998 046 88	0.001 953 12
15	1.999 938 97	0.000 061 03
20	1.999 998 09	0.000 001 91

Picturing a Number System

This splitting into groups of two, or its converse, combining in groups of two, is typical of the binary number system, just as the decimal system is based on splitting or combining in groups of ten. The fractal in Figure 1.2 is perhaps the most simple example of an extensive family of fractals in which the structure of a number system is represented geometrically. This gives us every reason to take a closer look at number systems. Anyone who wants to use computer programs to conjure up fractals on the screen will need to do this.

Numbers: Naming and Notation

To us counting goes almost without saying. We hardly stop to think that our present way of counting, both in words and notation, is the result of a long cultural-historical evolution. Its foundation was laid by the Hindus fourteen centuries ago and even earlier by the Chinese. Another ten centuries passed before Simon Stevin (1548–1620) from Bruges introduced modern decimal fractions in Europe.

It all seems so simple. Units are combined into units of a higher order, tens. In the same way tens are combined into hundreds, hundreds into thousands, etc. Neither words nor notation present any problems (for the time being):

a thousand	$=$	1 000	$=$	10^3
a million	$=$	1 000 000	$=$	10^6
a billion	$=$	1 000 000 000	$=$	10^9
a trillion	$=$	1 000 000 000 000	$=$	10^{12}

As we see, the notation in powers of ten is shorter and clearer, especially if then number of zeros is six or more. For typographical reasons the exponent is often written on the same line as the base, e.g., 10↑6 instead of 10^6, just like we do in computing. Unfortunately the correspondence of name and notation may differ from country to country. Thus the Dutch "miljard" is what Americans call a billion, 10↑9. To mathematicians this doesn't matter much since a number like 10↑100 is defined exactly by its notation. It isn't necessary to *say* it—in fact words fail here!

In decimal we can also define an unlimited number of ever smaller units: one tenth, one hundredth, one thousandth, etc. Every unit, whatever its order, is divided into ten smaller units of a lower order:

one tenth	=	$\frac{1}{10}$	=	0.1	=	10^{-1}
one hundredth	=	$\frac{1}{100}$	=	0.01	=	10^{-2}
one thousandth	=	$\frac{1}{1\,000}$	=	0.001	=	10^{-3}
one millionth	=	$\frac{1}{1\,000\,000}$	=	0.000\,001	=	10^{-6}

Here too notations like 1/100 or 10↑−2 can be used for typographical reasons.

The notation of any particular number, the present year for instance, can be found by splitting off powers of ten. We start with the biggest unit, one thousand. Of the remainder we can take nine hundreds, followed by nine tens, whereupon one unit is left. Thus

$$1991 \;=\; 1 \cdot 10^3 \;+\; 9 \cdot 10^2 \;+\; 9 \cdot 10^1 \;+\; 1.$$

For a number that isn't a whole number, π for instance (the ratio of a circle's circumference to its diameter), the result is a decimal fraction: $\pi = 3.141\,592\,653\,589\,79\ldots$ We can carry on with this indefinitely; it never stops. Even though in practice some ten places of decimals of π suffice for carrying out the most amazing astronomical calculations, the expansion of π as a decimal fraction carrying on forever is a continual challenge. Up to the fifteenth century the best approximations were $3.141\,592\,6 < \pi < 3.141\,592\,7$ and $355/113 = 3.141\,592\,9\ldots$ By A.D. 1700 a hundred decimal digits were known. Now, thanks to the computer, over 1011 million digits have been calculated!

The predominance of the *decimal* system is connected of course with the fact that we humans have ten fingers. It is likely that we are not the only intelligent form of life in the universe. Consequently, elsewhere a form of life using a different number system may have evolved—the *octal* (based on eight) by an octopus-like being, perhaps.

As a matter of fact, the decimal system has its drawbacks. Dividing something into three parts that are of exactly the same size takes some doing. To express 1/3 as a decimal fraction we have to make do with approximations:

$$0.3 \quad 0.33 \quad 0.333 \quad 0.3333, \quad \text{etc.}$$

This also goes on indefinitely, just like π, but here we *do* have a simple regularity in the process.

Five thousand years ago things were a lot better. In Mesopotamia the Sumerians developed a hexagesimal (60-based) system that was extremely satisfactory in practice (in agriculture and astrology, for instance). To them we still owe our division of time into hours, minutes, and seconds. And sure enough, one third of an hour is exactly twenty minutes.

The Egyptians, on the other hand, had a much more primitive notation based on powers of ten. They had individual symbols for 1, 10, 100, 1 000, 10 000 and 100 000. These were repeated as often as necessary. The well-known story of Atlantis mentioned by Plato in his *Critias* shows how complicated the Greeks thought the Egyptian system. They interpreted the Egyptian numbers as ten times too big—1 000 instead of 100, for instance. That is why Atlantis was so large that it didn't fit in the Mediterranean. In actual fact they were telling the true story of the gigantic volcanic eruption that destroyed the Minoan civilization on Crete.

Other peoples, the Maya for instance, developed the vigesimal system (20-based). Be that as it may, in our time the decimal system is supreme.

The Binary System

When dealing with problems of a mathematical or technical nature it is sometimes much more convenient to do calculations in other number systems. The most simple system is of course the binary one. Here we have to make groups of two all the time. If for the time being we confine ourselves to the natural numbers (0, 1, 2, 3, ...), it works like this:

decimal	0	1	2	3	4	5	6	7	8	9	10
binary	0	1	10	11	100	101	110	111	1000	1001	1010

We use the numbers 0 and 1 only. By the binary notation 110 100 111 we mean

$$110\,100\,111 = 2^8 + 2^7 + 2^5 + 2^2 + 2 + 1$$
$$= 256 + 128 + 32 + 4 + 2 + 1 = 423.$$

In binary the numbers are about three times as long as in decimal. This means more writing, but it makes calculating a lot easier. The multiplication tables for instance have shrunk to

$$0 \times 0 = 0, \qquad 0 \times 1 = 0$$
$$1 \times 0 = 0, \qquad 1 \times 1 = 1.$$

Internally computers calculate in binary; their speed is due to carrying out the same simple calculations a great many times.

From Decimal to Binary

The best method for deriving the binary expansion of a given natural number from its decimal form is to divide the number by two again and again, noting down the remainder each time. For example:

$$
\begin{aligned}
423 \div 2 &= 211 \quad &\text{remainder } 1 \\
211 \div 2 &= 105 \quad &\text{remainder } 1 \\
105 \div 2 &= 52 \quad &\text{remainder } 1 \\
52 \div 2 &= 26 \quad &\text{remainder } 0 \\
26 \div 2 &= 13 \quad &\text{remainder } 0 \\
13 \div 2 &= 6 \quad &\text{remainder } 1 \\
6 \div 2 &= 3 \quad &\text{remainder } 0 \\
3 \div 2 &= 1 \quad &\text{remainder } 1 \\
1 \div 2 &= 0 \quad &\text{remainder } 1
\end{aligned}
$$

In this algorithm successive digits in the binary expansion, reading from left to right, appear one above the other in the right-hand column reading from bottom to top. So 423 is 110 100 111 in binary.

Binary Fractions

In the same way as in the decimal system, we can express fractions and irrational (incommensurable) numbers in binary as a terminating or as an indefinitely continuing *binary* fraction. Thus

$$\frac{45}{64} = 0.101\ 101, \quad \text{i.e.,}$$

$$\frac{45}{64} = \frac{1}{2} + \frac{1}{8} + \frac{1}{16} + \frac{1}{64}.$$

Doubling, or multiplying by two, comes down to moving the point one position to the right:

$$\frac{45}{32} = 1.011\ 01,$$

$$\frac{45}{16} = 10.110\ 1, \quad \text{while}$$

$$45 = 101\ 101.$$

If one tries to express an arbitrary fraction p/q in binary, in which p and q are natural numbers with no common factor, one soon discovers that the expansion of the fraction will stop only if q is a power of two, as in the example just given. In all other cases it turns out that the same pattern keeps repeating itself in the expansion of the fraction. For example $3/7 = 0.011\ 011\ 011 \ldots$ or, in a shorter notation (underlining the repeating pattern), $3/7 = 0.\underline{011}$.

The algorithm for calculating the binary expansion is based on continually doubling the numerator of the fraction concerned:

0	1	1	0	1	1	0	...
3	6	12					
		5	10				
			3	6	12		
				5	10		
					3	6	...

The moment we find a number that is larger than or equal to the denominator (7, in our example), we subtract the denominator from it, note down a binary 1 above the line, and continue with the remainder. We will illustrate this once more, for the fraction $5/11$:

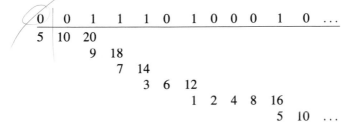

So $5/11 = 0.\underline{011\,101\,000\,1}$ with a recurring pattern of ten binary digits.

We noticed earlier, in connection with Figure 1.2, that the infinitely continuing series $1 + 1/2 + 1/4 + \ldots$ has limit 2. We can express this in binary as $2 = 1.111\,1\ldots$, a recurring fraction consisting of ones only. So in fact this is a somewhat laborious way of writing the number two! The notation $0.100\,111\,1\ldots$ or rather $0.100\underline{1}$ is less odd. But also in this case it is better to write this number as 0.101, or $5/8$ in decimal notation. Consequently we can agree to replace, where necessary, an infinite series of ones following a zero with a single *one* in the position of the zero. In decimal we have a similar situation: $0.999\,9\cdots = 1$ and $0.123\,\underline{9} = 0.124$.

The Quaternary and Octal Systems

Converting from binary to quaternary (base 4) or octal (base 8) is quite simple. It is similar to the transition from decimal to cental (base 100). For example:

$$
\begin{aligned}
423 \;=\;& 110\,100\,111 && \text{binary} \\
423 \;=\;& 1\,(10)\,(10)\,(01)\,(11) && \text{quaternary} \\
\;=\;& 1 \quad 2 \quad 2 \quad 1 \quad 3 && \text{quaternary} \\
\;=\;& 1 \cdot 4^4 + 2 \cdot 4^3 + 2 \cdot 4^2 + 1 \cdot 4^1 + 3 \\
\;=\;& 1 \cdot 256 + 2 \cdot 64 + 2 \cdot 16 + 1 \cdot 4 + 3
\end{aligned}
$$

$$423 = (110)(100)(111) \qquad \text{octal}$$
$$= 6 \quad 4 \quad 7 \qquad \text{octal}$$
$$= 6 \cdot 8^2 + 4 \cdot 8 + 7$$
$$= 6 \cdot 64 + 4 \cdot 8 + 7$$

The octal system frequently occurs in computer programs. It does not have the drawback of binary—such long numbers.

The transition from octal to binary presents no problems either. This time we will demonstrate it with $5/13$:

$$
\begin{array}{ccccc}
& 3 & 0 & 4 & 7 & \ldots \\
\hline
5 & 40 & & & \\
& 1 & 8 & 64 & \\
& & 12 & 96 & \\
& & & 5 & \ldots \\
\end{array}
$$

The algorithm gives the octal notation 0.<u>3047</u> for $5/13$. Conversion into binary is a matter of notation only: $3 = 011$, $0 = 000$, $4 = 100$, $7 = 111$. Hence $5/13$ is 0.<u>011 000 100 111</u> in binary.

The Ternary System

Finally, let us say something about ternary (base 3), which we will need later. Now,

$$423 = 1 \cdot 243 + 2 \cdot 81 + 2 \cdot 9$$
$$= 1 \cdot 3^5 + 2 \cdot 3^4 + 0 \cdot 3^3 + 2 \cdot 3^2 + 0 \cdot 3 + 0.$$

So 423 is 120 200 in ternary notation.

In ternary, only the numbers 0, 1, and 2 are used. In this system the multiplication tables are only slightly more complicated than in binary:

×	0	1	2
0	0	0	0
1	0	1	2
2	0	2	11

Neither does the expansion of a fraction present real problems. For the expansion of the fraction 3/13 (in decimal notation), we use the by now familiar algorithm:

$$\begin{array}{r|lll} & 0 & 2 & 0 \quad \dots \\ \hline 3 \;\; 9 & 27 & & \\ & 1 & 3 \quad \dots \end{array}$$

Hence 3/13 becomes 0.<u>020</u> in ternary.

"Ternary" Tree

The tree structure in Figure 1.3 is based on the ternary system. From one point, three identical main branches set off at angles of 120°. Each of the three ends is itself the starting point of three smaller branches in the same directions, and so it goes on.

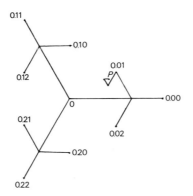

Figure 1.3 Diagram of the ternary tree

The direction to the right we mark with a 0, the one to the left upward with a 1, and the one to the left downward with a 2. Starting

from the original point 0, we can label the ends of successive branches
with ternary fractions:

first   0.0   0.1   0.2

then   0.00   0.10   0.20   (the next branches to the right),
        0.01   0.11   0.21   (the next branches to the left upward),
        0.02   0.12   0.22   (the next branches to the left downward),

and so on.

A couple of them are indicated in Figure 1.3. Thus point P in that
figure corresponds to the ternary fraction 0.012 10, to be read as right,
(left) upward, (left) downward, (left) upward, right. In this notation the
number of positions after the point (or the number of index numbers)
is fixed beforehand. This implies that we must supplement fractions
that terminate earlier with one or more extra zeros.

ON THE COMPUTER. In Figure 1.4 the tree has been traced further
on the computer. Here the branches are reduced five times by the same
factor (0.45 has been chosen), and the tree is said to have five *levels*.
The number of branches then totals 3↑6 or 729. This program is given
in Appendix B under the name TREE3.

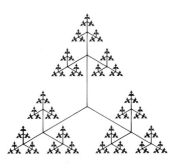

Figure 1.4 Ternary tree

Anyone who wants to run this program on a microcomputer has to take the resolution of the screen into account. This limits the number of levels, i.e., the order of approximation to the indefinitely continued ideal fractal. In the programs the order is usually indicated by p. For the ternary tree of Figure 1.4, $p = 5$.

For the more simple tree of Figure 1.2, we give a program called TREE2, with which seven levels can be attained. For the fractal shown in Figure 1.1 we have given two different programs, TREEH1 and TREEH2. The first program is fast and simple, but its execution takes up quite a lot of memory space. The second program requires much less memory space. It is based on the backtrack method, an efficient way of numbering the branches of a tree that has branched at least twice. The backtrack method will be discussed in Chapter 5.

Sierpinski's Sieve

In 1915 the Polish mathematician Vaclav Sierpinski (1882–1969) thought up a nice variation on the ternary tree. This is now known as Sierpinski's sieve. It is obtained by starting with an equilateral triangle thought of as a solid object. We divide this into four smaller equilateral triangles, of which the middle one is removed (Figure 1.5). This way, a triangular hole is produced. With the three remaining solid equilateral triangles we proceed in just the same way, so that three smaller triangular holes appear. In Figure 1.5 we carried out *one* more step, but in our mind we can of course carry on like this indefinitely.

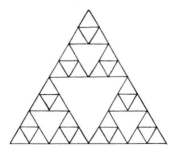

Figure 1.5 Diagram of Sierpinski's sieve

A much better drawing, made on a computer, is depicted in Figure 1.6. The program for this, SIER, is given in Appendix B.

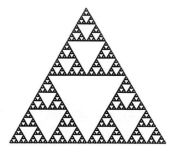

Figure 1.6 Sierpinski's sieve

CHAPTER 2

Numbers and Points

Mathematicians have always tried to base their line of reasoning, by strictly logical rules, on basic principles (axioms) that don't need to be proved themselves. About twenty centuries ago, the Greeks had already managed to work out such a system of axioms for geometry. One of their textbooks, the *Elements* by Euclid (330–275 B.C.), played a predominant part in mathematics well into this century. In education especially it served as a classic example of mathematical reasoning.

The Greeks thought very geometrically. A number immediately made them think of the length of a line-segment. Its second power they would relate to the area of a square; its third power, to the volume of a cube. They never got around to the power four! Irrational numbers like $\sqrt{2}$ gave them a lot of trouble. It is amazing that a mathematically rigorous theory in which the irrational number has a logical place evolved only in the last century. This we owe chiefly to the German mathematician Georg Cantor (1845–1918). In this chapter we discuss a few of his ideas, as they are of great importance to the mathematical background of fractals.

Cantor Fractal

Cantor is one of the founders of set theory (as it is called today). To us, however, Cantor's chief importance lies in the fact that he thought up something one could call the oldest fractal. This so-called *Cantor point-set*, dating from 1870, can be built up by repeating a simple principle indefinitely.

We start with a line-segment, including its endpoints, as shown in Figure 2.1. Of this we leave out the middle third, but not the endpoints. We are then left with two line-segments with a total of four endpoints.

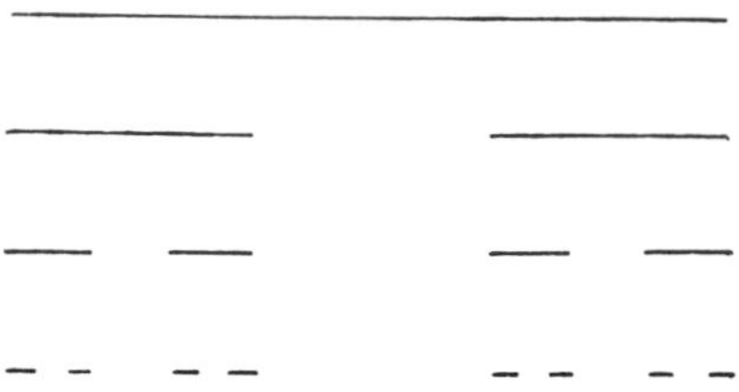

Figure 2.1 Construction of Cantor's fractal

We treat each of these two line-segments like the original one: the middle thirds are removed, so that a total of four line-segments with two endpoints each remain. We continue like this. Eventually we will be left with discrete (i.e., separate or individual) points only: the Cantor point-set or fractal. Such a fractal, built up from discrete points, is now usually called "dust"—in our case, "Cantor dust."

If we make the length of the original line-segment 1, then after three steps there will be $2 \uparrow 3 = 8$ line-segments, each of length $3 \uparrow (-3) = 1/27$. After n steps that makes $2 \uparrow n$ line-segments, each of length $3 \uparrow (-n)$. The total length of the remaining line-segments is $(2/3) \uparrow n$. This tends to zero as n increases indefinitely.

A mathematician would say that Cantor's point-set has measure or dimension zero. In Chapter 5 we will learn that there is another, more practical concept of dimension, according to which the Cantor set has dimension $0.6309\ldots$. This dimension is not a whole number, but a fraction. Hence the term "fractal dimension."

Cantor's Comb

Making a reasonably good picture of the Cantor fractal is not easy. However, by using a nice trick we can generate a comblike structure, the Cantor comb (Figure 2.2, made with the program COMB). We begin with a horizontal bar of length 1. At each stage in the development of the Cantor set shown in Figure 2.1, we turn the new line-segments into teeth of equal height and join them onto the comb. Of course Figure 2.2 shows five stages and only gives us an approximation. In actual fact the "teeth" of the comb keep on going down indefinitely.

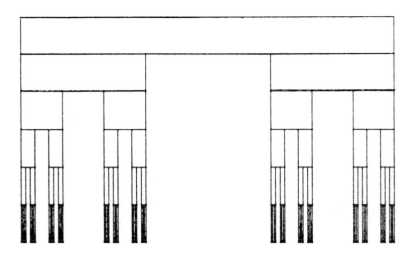

Figure 2.2 Cantor's comb

Points as Numbers, Numbers as Points

In the seventeenth century, new views on geometry started catching on. Calculating skill had evolved quite far, and people started thinking of converting geometrical processes into arithmetical ones. René Descartes (1596–1650), a French philosopher, was the pioneer of this field. He joined the army of Prince Maurice of Nassau (Prince of Orange) and stayed in Holland for some time.

The location of a point in a plane can be fixed by two (real) numbers—the distances (with a plus or minus sign) from that point to two perpendicular lines, the coordinate axes. These numbers are called Cartesian coordinates, Cartesius being the Latin version of Descartes. We use a horizontal coordinate axis, the X-axis, and a vertical one, the Y-axis, as shown in Figure 2.3. Their intersection O is called the origin. In Figure 2.3 the coordinates of point P are indicated by the letters x and y. The notation (x, y) can therefore be interpreted either as two real numbers, or as the coordinates of a point.

If we restrict ourselves to points on the X-axis, just their x-coordinate will fix them, because on the X-axis $y = 0$. Each point on the X-axis

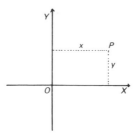

Figure 2.3 Cartesian coordinates

corresponds to a number, the x-coordinate, and conversely each real
number corresponds to a point. Thus the X-axis is a number line, a
measuring rod of infinite length. We take from this the part corresponding to the numbers between 0 and 1, the line-segment OE. On
it O, the origin, is the endpoint on the left, and E corresponds to 1.
If P is the point that corresponds to the number a, then $OP : OE = a$
(Figure 2.4).

Figure 2.4 Number as ratio between line-segments

This way we can convert all we say about numbers into corresponding statements about points, and vice versa. If, for instance, P and
Q are the image points of the numbers a and b, then the midpoint of
the line-segment PQ corresponds to the arithmetical mean or average,
$(a + b)/2$, of the two numbers.

This shuttling back and forth between arithmetical and geometrical
concepts is now second nature to mathematicians. Computer program-
mers designing graphics programs no doubt experience this too. After
all, everything drawn by the computer has to be expressed in Cartesian
coordinates first.

The Cantor Fractal, Seen Arithmetically

We can now also express the construction of a Cantor fractal arithmetically. As line-segments are continually divided into three equal parts, it is most convenient to use the ternary system. Because we are dealing with numbers beween 0 and 1, this involves expanding fractions like this:

$$a = \frac{c_1}{3} + \frac{c_2}{9} + \frac{c_3}{27} + \frac{c_4}{81} + \cdots$$

or $a = 0.c_1c_2c_3c_4\ldots$, in which c_1, c_2, c_3, $\ldots$ can be the numbers 0, 1, or 2. For instance,

$$0.3 = \frac{3}{10} = 0.\underline{0220}$$

$$0.5 = \frac{1}{2} = 0.\underline{1}$$

$$0.8 = \frac{4}{5} = 0.\underline{2101}$$

By underlining part of the fraction, we indicate that a group of digits has to be repeated again and again. This phenomenon is typical of common (commensurable) fractions, not only in ternary, but also in any other number system.

The construction starts with the interval [0,1], including its endpoints 0 and 1. The middle third (i.e., all numbers that have a ternary fraction starting with 1) is marked. At the next step we do the same for the second position after the point (Figure 2.5). In the end it comes down to removing all numbers that have a 1 in their expansion as a ternary fraction.

The diagram of Figure 2.5 shows that a lot of numbers disappear at the first stroke—the number 0.5 (in decimal), for instance. The point 0.8 disappears at the second stroke, since $0.8 = 0.\underline{2101}$ (ternary), as we saw earlier. The number $0.3 = 0.\underline{0220}$ (ternary), however, never goes. As a result, the points of the Cantor set can be identified as all the numbers between 0 and 1 that can be written in ternary with zeros and twos only! Of course 0 is included, but also 1, since $1 = 0.\underline{2}$ (ternary).

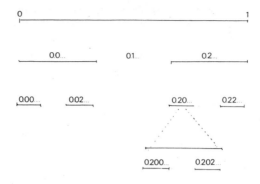

Figure 2.5 Cantor fractal and the ternary system

Cantor's point-set is special in more ways than one. Before we can discuss this, however, our basic mathematical notions have to be extended. In the following passages we will explain some of the more fundamental mathematical aspects of fractals. We shall be staying close to the theories Cantor himself developed.

Different Kinds of Infinity

Imagine for a moment that we are back in an early primitive culture. In this culture cattle-breeding has progressed to such an extent that the nomads have very large herds. Unfortunately they do not have a good counting system. Suppose they cannot do much more than using their hands and fingers for counting. Still, they can make accurate statements like "I have as many cows as sheep" or "I have as many cows as my friend has horses." To check this they simply drive all the cows and horses together and then keep releasing a cow *and* a horse at the same time, until there are no animals left. Mathematicians use a very abstract term for this: they talk of a one-to-one correspondence between the elements of two sets.

BACK TO THE ORIGINS. Two sets have the same size, or are equivalent, if one can exactly match the elements of one set with those of the other. In our example, for instance, each horse matches one cow; conversely, each cow matches one horse. Nowadays cattle-breeders simply note down the number of their stock in modern number-notation. Mathematicians, however, are sometimes forced to return to the origins of culture.

For example, what do you make of the statement *There are as many even numbers as odd numbers*? There are infinitely many of each sort, but we cannot use infinity (indicated by ∞) in our calculations. What we *can* do is to match the odd and even numbers just like our prehistoric stockman did:

$$
\begin{array}{cccccc}
1 & 3 & 5 & 7 & 9 & \ldots \\
2 & 4 & 6 & 8 & 10 & \ldots
\end{array}
$$

This works very well. Every number gets its turn and has its match; the statement now makes sense.

DENUMERABLE INFINITY. Now for the next statement: *The number of squares equals the number of all the natural numbers*. This seems odd, because as we continue in the series of the natural numbers

$$\underline{0} \quad \underline{1} \quad 2 \quad 3 \quad \underline{4} \quad 5 \quad 6 \quad 7 \quad 8 \quad \underline{9} \quad 10 \quad 11 \quad 12 \quad 13 \quad 14 \quad 15 \quad \underline{16} \quad \ldots$$

squares (underlined) occur less and less often. Still, the infiniteness of the number of squares is of the same type as that of all the natural numbers. It so happens that the matching principle works without problems:

$$
\begin{array}{ccccccccccc}
0 & 1 & 2 & 3 & 4 & 5 & 6 & 7 & 8 & 9 & 10 & \ldots \\
0 & 1 & 2^2 & 3^2 & 4^2 & 5^2 & 6^2 & 7^2 & 8^2 & 9^2 & 10^2 & \ldots
\end{array}
$$

This kind of infinite set is called *denumerably* (countably) *infinite*. In the same way, it can be shown that the set of all third powers, or any higher power, is also denumerable.

As we said, Cantor was the originator of these reflections on "infinitely many." He also showed that there exist sets with a number of elements that, though infinite, is *not* denumerably infinite. We will encounter his ideas frequently.

Hilbert Hotel

David Hilbert (1862–1943), considered to be one of the foremost mathematicians at the beginning of this century, used the following story to
illustrate the concept "denumerably infinite."

A hotel has a denumerably infinite number of rooms, numbered 1,
2, 3, ... All rooms are occupied. A new guest arrives. The porter,
not being a mathematician, tells him the hotel is fully booked and
that there is no room. The manager, however, knows how to handle
the situation. He makes all guests shift one position, so that room 1
becomes available for the new guest.

Next, a queue of a hundred new guests arrives.

"Fully booked," says the porter.

"Don't worry," the manager says; "Just move the guests of rooms 1,
2, 3, . . . to rooms 101, 102, 103, ..." This way a hundred rooms
become available, and everyone is content.

Finally, a queue of a denumerably infinite number of new guests
arrives, a queue of truly infinite length. The porter is desperate, but
once again the manager finds a way out. He makes each guest move
to a room with a number double his previous room number. So:

$$1 \rightarrow 2 \quad 2 \rightarrow 4 \quad 3 \rightarrow 6 \quad \ldots$$

This way all rooms with odd numbers become vacant and can, one
after the other, be allocated to a new guest. Everyone gets his turn.
Once again everybody is happy.

As Many Fractions as Natural Numbers

The following statement is even more surprising: *There is the same
number of fractions as natural numbers.* By fractions we mean real
fractions like p/q, p and q being natural numbers (q must not be
zero). We assume the fraction has been reduced as much as possible.
For instance, 24/60, 12/30, and 8/20 will not occur, as they can be
reduced to 2/5. On the face of it, thinking up a suitable matching
seems quite difficult. We cannot enumerate the fractions by ranging
them according to increasing size, even if we restrict ourselves to the
fractions between 0 and 1. There *is* no smallest fraction after zero. We

simply cannot get away from zero. The following list shows us how to deal with this problem:

$$\frac{1}{1} \quad \frac{1}{2} \quad \frac{1}{3} \quad \frac{2}{3} \quad \frac{1}{4} \quad \frac{3}{4} \quad \frac{1}{5} \quad \frac{2}{5} \quad \frac{3}{5} \quad \frac{4}{5} \quad \frac{1}{6} \quad \frac{5}{6} \quad \cdots$$

Here the fractions are arranged first by increasing denominator and then by increasing numerator. As we see, each fraction has its turn in this enumeration. This means that the set of fractions between 0 and 1 is denumerably infinite.

In order to enumerate the fractions bigger than one, we use the last trick used by the Hilbert hotel (moving to rooms with double the number). First, here is the list of new guests:

$$\frac{1}{1} \quad \frac{2}{1} \quad \frac{3}{1} \quad \frac{3}{2} \quad \frac{4}{1} \quad \frac{4}{3} \quad \frac{5}{1} \quad \frac{5}{2} \quad \frac{5}{3} \quad \frac{5}{4} \quad \frac{6}{1} \quad \frac{6}{5} \quad \cdots$$

This is obtained by inverting each fraction from the first series. If we combine both series,

$$\frac{1}{1} \quad \frac{1}{2} \quad \frac{2}{1} \quad \frac{1}{3} \quad \frac{3}{1} \quad \frac{2}{3} \quad \frac{3}{2} \quad \frac{1}{4} \quad \frac{4}{1} \quad \frac{3}{4} \quad \frac{4}{3} \quad \frac{1}{5} \quad \cdots$$

then we get an enumeration of all (positive) fractions. The natural numbers are included by interpreting them as fractions with denominator 1.

Irrational Numbers

Fractions are called rational or commensurable numbers, because they can be measured as the ratio of two natural numbers. In addition to these there are irrational or incommensurable numbers, like π or $\sqrt{2}$. These numbers can be expressed as infinite decimals. Their expansion neither terminates nor is periodic. It does not make any difference which number system is used. In decimal or in any other number system, the expansion of an irrational number is irregular and continues indefinitely. For the sake of convenience we will stick to decimal.

Every terminating or periodic decimal fraction supplies us with a rational number. Any fraction that continues indefinitely, without a periodic pattern, will always be an irrational number. If we terminate this expansion somewhere, the resulting approximation is a rational

number. The accuracy, i.e., the error made, depends on where exactly
we terminate. We can make the error as small as we like. In that sense
an irrational number can always be interpreted as the limit of a series
of rational numbers that approximate it better and better.

Indenumerably Infinite

If we think of the rational numbers between 0 and 1 arranged by size,
the interval will seem to be completely filled by them. Between two
rational numbers, however close together, there will always be another
rational number—an infinite number of them, in fact. To appreciate
this, we need consider only the (arithmetic) mean. Applying this to the
fractions $5/12$ (= $0.416\,666\ldots$) and $3/7$ (= $0.428\,571\ldots$), we get
$71/168 = 0.422\,619\ldots$.

This can be done more elegantly, however. We make a new fraction
by adding the denominators and the numerators separately:

$$\frac{5+3}{12+7} = \frac{8}{19} = 0.421\,053\ldots$$

It looks as if the rational numbers fill the interval $[0,1]$ completely.
All the same, irrational numbers are between them everywhere. Even
more odd is the fact that there are "many more" irrational numbers
than rational ones. In Cantor's own words: *The set of the irrational
numbers is indenumerably infinite.*

With Cantor's statement we encounter a second kind of infinity.
Proving the statement seems difficult at first, but Cantor's reasoning is
so original and so simple that we will take a look at it. It is an example
of a so-called "proof by contradiction." We start with the opposite of
what we want to prove and show that this leads to an inconsistency.

With Cantor let us assume that all numbers between 0 and 1, rational
and irrational ones together, are denumerable. We could then list them
as a_1, a_2, a_3, $\ldots$. We imagine each one of these numbers written
as a decimal fraction that continues indefinitely. If the fraction has a
finite number of decimal digits, we simply continue the expansion with
zeros.

Next we construct a new number b, digit by digit. In doing this we
arrange that the first digit of b differs from the first digit of a_1, the

second digit of b differs from the second digit of a_2, the third digit of b differs from the third digit of a_3, etc. The number b built up in this way ought to occur in the list too, as the number a with index n, say. But owing to the construction, b and a differ in their nth decimal digit. Consequently a contradiction has arisen. This can be resolved only by rejecting the basic assumption, i.e., that of denumerability.

Limit Points

When describing fractals we often use the term *limit* or *limit point*. The idea here is that we have an infinite sequence of numbers, a_1, a_2, a_3, ..., say, that approaches a certain number, the limit. A few simple examples:

$$1 \quad \frac{1}{2} \quad \frac{1}{2^2} \quad \frac{1}{2^3} \quad \frac{1}{2^4} \quad \cdots \quad \to 0$$

$$\frac{1}{2} \quad \frac{2}{3} \quad \frac{3}{4} \quad \frac{4}{5} \quad \frac{5}{6} \quad \cdots \quad \to 1$$

$$3 \quad 3.1 \quad 3.14 \quad 3.141 \quad 3.1415 \quad \cdots \quad \to \pi$$

By saying L is the limit of the sequence a_n, where $n = 1, 2, 3, \ldots$, we mean that eventually, when the index n is large enough, the deviation of a_n from the limit becomes arbitrarily small. In mathematical language:

$$|a_n - L| \to 0 \qquad \text{as } n \to \infty.$$

The third example demonstrates that an irrational number such as π is the limit of a sequence of rational numbers. These numbers can be obtained from the decimal expansion of π by terminating the expansion one position further each time. The error we make is smaller than the last decimal unit: 1, $1/10$, $1/100$, $1/1\,000$, $1/10\,000$, ... in turn. Similarly, every irrational number can be thought of as the limit of a sequence of rational numbers.

Let us also express this geometrically. If we look at the line-segment OE representing the numbers between 0 and 1, then the rational numbers, the proper fractions, will correspond to rational points. There are a denumerably infinite number of these. They form a dustlike point-set full of "microscopic" holes, since in between two rational numbers there are always irrational ones. These "holes," the irrational (or incommensurable) numbers or points, are the limits—or rather, in geometrical language, the limit points—of sequences of rational points.

We now reach an intuitively rather paradoxical conclusion. The rational points between 0 and 1 constitute a denumerable set. The "holes" in that set, the irrational points, are *not* denumerable, but form a set of a higher order of infinity!

The concept of a limit can easily be extended to point-sets in the plane, and as such can be applied to all the fractals we meet in this book. A recurring situation is that of a fractal being built up step by step from point-sequences. After an infinite number of steps, a point-set with a denumerably infinite number of points will then have been formed. The limit points are then added to these, i.e., the holes are filled in. This can have varying results. A dust fractal consisting of infinitely many points disconnected from one another, like the Cantor set (Figure 2.1), may have been formed. On the other hand "filling in the holes" may also lead to cohesion, like the Koch fractal (Figure 3.1), a continuous meandering line.

Finally, it is possible that adding the limit points produces a plane-filling fractal. We will encounter this situation when we discuss the dragon curve (Figure 3.24) at infinite resolution. It is easy to think up an even more simple example of such a situation. We first form the set of all rational points (x, y) in the unit square, i.e., all points for which both x and y are proper fractions. That set consists of a denumerably infinite number of disconnected points. Addition of the limit points results in *all* points of the square with $0 \leq x \leq 1$, $0 \leq y \leq 1$.

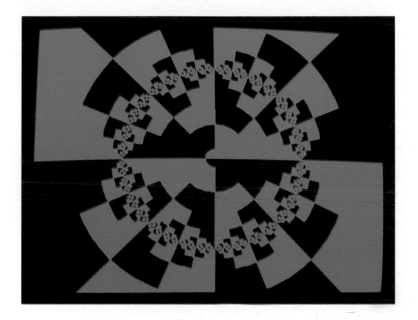

Binary division of a circular region

CHAPTER 3

Meanders and Fractals

In this chapter we introduce the concept "fractal" by means of an experiment that is extremely relevant to our subject. The object of the experiment is to determine the length of the British coastline. It leads to a measure of the degree of meandering of a coastline or national border. Later Benoit B. Mandelbrot called this measure, a number between 1 and 2, the "fractal dimension."

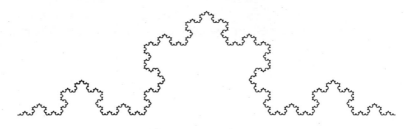

Figure 3.1 Koch fractal

After we have discussed that experiment, we will look at a fractal that is most interesting historically and was thought up by the mathematician Helge von Koch in 1904. It is a mathematically defined coastline with triangular promontories whatever the scale of magnification (Figure 3.1). It is an example of a curve that does not have a tangent anywhere. This fractal is closely related to the Cantor point-set we discussed in the previous chapter. The Koch fractal is the prototype of an extensive family of fractals based on the repetition of a simple geometrical transformation. Under this transformation, a line-segment (the base) is replaced by a bent line (the motif). We will give a number of examples of fractals of this family; we hope the reader will carry out further experiments himself with pencil and paper, or—better still—a microcomputer. A fairly general program for this is included in Appendix B.

Richardson's Experiment

Nowadays, thanks to the publications of the mathematician Benoit B. Mandelbrot, fractals receive a lot of attention. In his beautiful book *The Fractal Geometry of Nature*, we can find much on the subject, along with lovely illustrations, historical anecdotes, and so on. The work demands a lot of the reader. The experiment of Lewis Fry Richardson (1881–1953), a somewhat eccentric English meteorologist, may have persuaded or at least stimulated Mandelbrot to make fractals his life's work.

In the work of Richardson published after his death, Mandelbrot in 1961 came across measurements for determining, among other things, the length of Britain's west coast and the Spanish-Portuguese land frontier. Richardson had noticed that the results depended heavily on the scale of the map used. A map with 1 cm corresponding to 100 km (scale 1 : 10 000 000) simply shows less detail than a hiking map, in which 1 cm corresponds to 1 km (scale 1 : 100 000). As we see more detail, the coastline gets longer!

The same phenomenon can also be expressed by considering one map only, on which all detail can be seen, but using a smaller measuring unit each time. We could make a first rough estimate with a measuring unit of 100 km, and then scale this down by a factor of 10 each time. Of course the choice of the factor 10 is rather arbitrary. The measuring unit could be reduced just as well by a factor of 2 or 3 each time.

Figure 3.2 shows us what Richardson actually found. On it the measuring unit is plotted horizontally on a logarithmic scale. We'll indicate it by a. As we move to the left a decreases. The measured length s has been plotted vertically, also on a logarithmic scale. The resulting lengths are indicated by small circles.

The fact that the results turn out to be roughly on a straight line is very remarkable. For the west coast of Britain this can be expressed by the formula

$$\log s = -0.22 \log a + \log s_1,$$

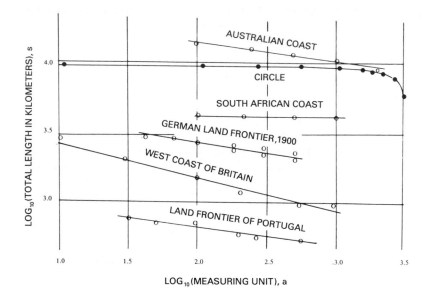

Figure 3.2 Results of Richardson's experiment

s_1 being the length when using a measuring unit of 1 km. The number 0.22 measures the slope of the line of results.

The formula defines a linear relationship between the logarithms of the measuring unit and the resulting length of the coastline. Without logarithms, this can be expressed as

$$s = s_1 a^{-0.22}$$

or as

$$s = s_1 \left(\frac{1}{a} \right)^{0.22}$$

This shows us what happens when we reduce a. If a is reduced by a factor of 32, s will double. In our mind we can continue reducing further and further. Eventually this would lead us to conclude that the west coast of Britain is infinitely long!

Of course, this is true only if the meandering of the coastline goes on repeating itself on an ever diminishing scale. To a certain extent this is true, as we could imagine a surveyor walking the length of the coast again and again using a shorter yardstick each time. He will measure rocks that are large in relation to his yardstick quite carefully, whereas he will disregard rocks that are relatively small.

Apparently the concept of *length* of a coast or boundary is not very practical. It would be better to specify the nature of such a boundary, the degree of meandering, by a number. For this we could take the slopes in Richardson's graph (Figure 3.2). As we saw, for the west coast of Britain this is about 0.22. For a line that is mathematically smooth, however, a circle for instance, successive scaling down eventually has no effect whatsoever. There the line of resulting lengths would be horizontal—i.e., with slope 0. Mandelbrot adds 1 to the slope, getting a number D, which he calls the "fractal dimension" of that coastline or frontier.

Degree of Meandering

The concept of dimension has various mathematical connotations, the most common of which is known as the topological dimension. This is usually an integer (whole number). Thus point, line, and plane have dimension 0, 1, and 2, respectively, and we live in a 3-dimensional space. Giving this concept of dimension a sound mathematical basis is far from easy. The Dutchman L.E.J. Brouwer (1882–1966), one of the major mathematicians of the twentieth century, led the way here. Other mathematicians, such as Felix Hausdorff (1886–1942), A. S. Besicovich (1891–1970), and A. N. Kolmogorov (1903–1987), defined it differently. Their definitions do not necessarily lead to integer dimensions. Applied to a meandering line, one of their definitions comes down to the following.

We select an arbitrarily small measurement unit a, the yardstick. Next we measure the length of the meandering line by approximating it as closely as possible with a bent line made up of equal line-segments of length a. If we suppose the yardstick is used N times, so that the total length measured is Na, then according to Mandelbrot's definition the "fractal dimension" is given by

(A) $\qquad D = \lim_{a \to 0} \dfrac{\log N}{\log\left(\frac{1}{a}\right)}.$

We will call D the "degree of meandering" of the meandering line for the moment; we will introduce a better and more general concept of dimension later. Note that in this formula $\log(1/a)$ equals $-\log a$. What is more, it does not matter whether we use logarithms to base 10 or to any other base. In applications we often reduce a in small steps, by a factor of 3 each time, for instance. The fraction $\log N / \log(1/a)$ then approaches a fixed value, the limit D. In some cases the fraction has the same value at every step. We can then write formula (A) more simply as

(B) $\qquad D = \dfrac{\log N}{\log\left(\frac{1}{a}\right)},$

or, equivalently,

$$N = \left(\frac{1}{a}\right)^{D}$$

If we recall that Na is the total length measured, which before we wrote as s, we then have the final formula:

$$s = \left(\frac{1}{a}\right)^{D-1},$$

which shows clearly once again how the length measured increases as the measuring unit decreases.

Koch Curve

In 1904 the mathematician Helge von Koch gave an example of a curve without a tangent anywhere. To his contemporaries this was a shattering experience, a curiosity without practical value, a figment of the imagination. Just imagine, a curve consisting of components each one of which, however small, is infinitely long! Now, nearly a century later, such "pathological curves" (as they were called) turn out to occur everywhere in pure and applied mathematics. Figure 3.1 is a good approximation to von Koch's curve. However, the actual curve exists only in our mind. Tracing it with an infinitely thin pencil would take an infinite amount of time, even if we ran along the curve at the speed of light.

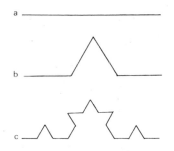

Figure 3.3 Construction of the Koch fractal

We can picture von Koch's curve in our mind as the limit of a series of approximations as shown in Figure 3.3. The construction looks very similar to that of the Cantor point-set of the previous chapter. Here too we start off with a line-segment, the base, and remove the middle third. Now, however, we fill the gap with the upright sides of an equilateral triangle. In this way a bent line consisting of four equal line-segments [Figure 3.3 (b)], the so-called motif, results. If the base has length 1, then the motif will consist of four line-segments, of length 1/3 each. Consequently the total length will be 4/3.

In the next phase, each one of the four line-segments is taken as a base and replaced by the corresponding scaled-down motif. The result will be a bent line consisting of $4 \times 4 = 16$ line-segments [Figure 3.3(c)] and having a total length of 16/9 or $(4/3)^2$. In the same way the third step produces $4^3 = 64$ line-segments of length $(1/3)^3 = 1/27$ each.

Since in this case

$$\frac{\log 4}{\log 3}, \quad \frac{\log 4^2}{\log 3^2}, \quad \frac{\log 4^3}{\log 3^3}, \ldots$$

is always the same, we can apply formula (B). We then find

$$D = \frac{\log 4}{\log 3} = 1.26\ldots$$

In the limit, all middle thirds of the original base [Figure 3.3(a)] have repeatedly been taken away. What is left is exactly the Cantor set. Von Koch's curve is self-similar. Every part, however small, is itself a miniature copy of the whole, so it is full of Cantor point-sets.

BETTER UNDERSTANDING. There are different ways of writing programs for the Koch curve. Let us describe the mathematical background of the program KOCH, which is given in Appendix B. This will give us a better understanding of the structure of the curve and of fractals in general.

First we fix the level of approximation. We'll denote it by the number p, the *order*. This means we will apply p transformations to the base line-segment. The result will be a meandering line consisting of $4 \uparrow p$ line-segments of equal length. Later, when we run the program on the computer, we can make p equal to 4 or 5, for instance.

We think of the line-segments as being numbered from 0 up to and including $(4 \uparrow p) - 1$. For every index number n, therefore, a line-segment, or rather a vector, of length $(1/3) \uparrow p$ has to be drawn. This comes down to determining the vector's direction. This is done as follows.

Write down the index n of the line-segment in quaternary (base 4). For example, for line-segment 482 in the meandering line of order $p = 5$ we get

$$
\begin{aligned}
482 \quad &= \quad 1 \times 256 \quad + \quad 3 \times 64 \quad + \quad 2 \times 16 \quad + \quad 0 \times 4 \quad + \quad 2 \\
&= \quad 1 \qquad\qquad 3 \qquad\qquad 2 \qquad\qquad 0 \qquad\quad 2
\end{aligned}
$$

The four possible directions (actually, two of them are the same) are each given a number we read from the motif (Figure 3.4). Adding the directions corresponding to the numbers in the quaternary expansion, we find that the direction of line-segment 482 is

$$\phi \quad = \quad \left(\tfrac{\pi}{3}\right) \quad + \quad 0 \quad + \quad \left(-\tfrac{\pi}{3}\right) \quad + \quad 0 \quad + \quad \left(-\tfrac{\pi}{3}\right) \quad = \quad -\tfrac{\pi}{3}\,.$$

Figure 3.4 Motif of the Koch fractal

The general formula is

$$n = t_0 + t_1 \cdot 4 + t_2 \cdot 4^2 + \cdots + t_{p-1} \cdot 4^{p-1}$$
$$\phi = a(t_0) + a(t_1) + a(t_2) + \cdots + a(t_{p-1}),$$

where $a(0) = a(3) = 0$, $a(1) = \frac{\pi}{3}$, and $a(2) = -\frac{\pi}{3}$. It is not hard to understand why this works. To explain this we will show, using Figure 3.5, how line-segment 482 evolves from the original base. Each of its five quaternary digits selects one of the four possible directions. To make this clearer we have omitted the scale reductions in Figure 3.5.

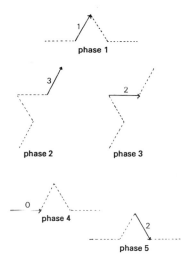

Figure 3.5 Koch fractal and the quaternary system

In fact, it would be better not to use integers to number the line-segments of the pth approximation, but rather to use quaternary fractions that have p digits after the point. To make this change we only have to divide by $4 \uparrow p$. In our example, with $p = 5$, this results in

$$\frac{482}{1024} = 0.13202.$$

If we imagine p infinitely large, each line-segment will be reduced to a point. Each point corresponds to an indefinitely continuing quaternary fraction, and vice versa. These fractions are all numbers between 0 and 1, both rational and irrational ones, uncountably many. Consequently we conclude that von Koch's curve is a true copy—in mathematical terms, a continuous image—of the unit line-segment (the base), i.e., of all numbers between 0 and 1.

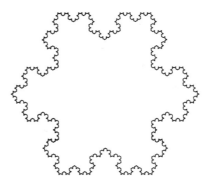

Figure 3.6 Koch island

VARIATIONS. If we place a fractal built up from line-segments, like von Koch's curve, on all sides of a regular polygon, nice pictures result. If, for instance, we take each of the three sides of an equilateral triangle as the base of a Koch curve facing outward, the result will be Figure 3.6. Mandelbrot called this a Koch island. Of course von Koch's curves could also be placed facing inward. In that case we get something completely different, Figure 3.7.

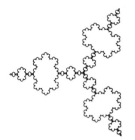

Figure 3.7 Koch island, inside out

The Repeated Motif

Von Koch's curve is the prototype of a large group of fractals. We start with a line-segment (the base) and a motif, a bent line-segment that at each step replaces each straight line-segment. In the Koch curve this is the step from (a) to (b) in Figure 3.3.

Another example of this is a motif consisting of eight equal line-segments (Figure 3.8). After four steps this already results in the meandering line of Figure 3.9, the so-called Minkowski sausage. Mandelbrot gave it this name to honor the friend and colleague of Einstein who died so untimely (1864–1909).

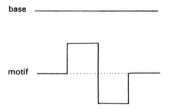

Figure 3.8 Motif for Minkowski fractal

Here, too, formula (B) with $N = 8$ and $a = 1/4$ can be applied at the first step. The result is $D = 1.5$. The program for tracing the Minkowski sausage is included in Appendix B under the name MINK. This program is very similar to that for the Koch curve, except that we have to use octal.

Figure 3.9 Minkowski fractal

GENERAL PROGRAM. MEANDER is a general program that enables us to use an arbitrary motif (provided it is not made up of too many lines). The fractal that results from this can be placed on an arbitrary initial line (or lines)—on the sides of an equilateral triangle as in Figures 3.6 and 3.7, for instance. In this general program we assume that the base consists of u line-segments and the motif of v line-segments. The vertices for both base and motif have to be listed in separate data files. For this we will use the basic principles of coordinate geometry as discussed in the previous chapter.

In Figure 3.10, O is the origin with coordinates $(0,0)$ and E is the unit point with coordinates $(1,0)$. The coordinates of an arbitrary point are the distances (with a plus or minus sign) parallel to the X-axis and the Y-axis.

A very simple motif of three equal line-segments, perpendicular to one another, is shown in Figure 3.10. This motif is fixed by the points O, P_1, P_2, and E; P_1 has coordinates $(0.4, 0.2)$ and P_2 $(0.6, -0.2)$. This makes $OP_1 = P_1P_2 = P_2E = 1/\sqrt{5}$.

For the base we can take a square, say. To do this the coordinates of the vertices have to be included in the data file of the program: $(1,1)$,

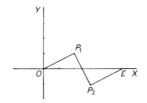

Figure 3.10 Motif in Cartesian coordinates

$(-1, 1)$, $(-1, -1)$, $(1, -1)$, $(1, 1)$. As the vertex $(1, 1)$ is both initial point and endpoint, it is listed twice.

After entering these data in the program as

$u = 4$, $v = 3$

data baseline 1, 1, -1, 1, -1, -1, 1, -1, 1, 1

data motif .4, .2, .6, $-.2$

we still have to fix the order of the approximation, p. It will depend on the available memory space and on the resolution of screen or plotter, as the case may be.

In general, $(v \uparrow p) - 1$ new vertices will be generated from the line-segment OE. If we reserve $2 \uparrow 10 = 1024$ memory locations each for the x- and the y-coordinate of each vertex, as in the version given here, we can go as far as $p = 6$ if $v = 3$. While calculating the coordinates of the $(v \uparrow p) + 1$ vertices of the meandering line generated from OE, we arrange that they conveniently end up in an array $x(n), y(n)$. The index n of this array runs from 0 to $v \uparrow p$. When the calculation is over, these coordinates can be used for each side of the base.

In the program the coordinates of the vertices are calculated step by step by repeatedly replacing a small line-segment by a reduced copy of the original motif, just as we did when we built it up geometrically. Mathematically this comes down to carrying out a similarity transformation, the essence of our computer programs. We will return to this subject in greater detail later. Here we will restrict ourselves to giving the formulas for the transformation illustrated in Figure 3.11. We suppose the motif OPE is to be transformed *similarly* to the new base $O'E'$, O' and E' being given by their coordinates (x_1, y_1) and (x_2, y_2), respectively. In other words: $(0,0) \rightarrow (x_1, y_1)$ and $(1,0) \rightarrow (x_2, y_2)$.

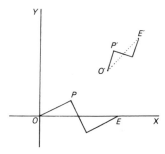

Figure 3.11 Similarity transformation

If the vertex P of the motif has coordinates (x, y), then those of the transformed vertex P', (x', y') say, are given by

$$x' = (x_2 - x_1)x - (y_2 - y_1)y + x_1$$
$$y' = (y_2 - y_1)x + (x_2 - x_1)y + y_1.$$

Experimenting

After this mathematical interlude, necessary if we want to understand the computer programs, we can experiment to our heart's content. A number of possibilities are summarized below. In each case we give the data for the base, the motif, and so on.

The first experiment was based on Figure 3.10, with a square as base. In the series of examples that follow we will take a line-segment, a square, or a triangle as initial line or base. When doing this the orientation matters, as the meandering line can be placed on either side of a straight line-segment. This results in the following possibilities.

Line-segment, single
coordinates (0,0), (1,0)
$u = 1$

Line-segment, double
coordinates (0,0), (1,0), (0,0)
$u = 2$

Square, inward
coordinates $(-1,-1)$,
$(1,-1)$, $(1,1)$, $(-1,1)$,
$(-1,-1)$
$u = 4$

Square, inward
coordinates $(-1,0)$,
$(0,-1)$, $(1,0)$, $(0,1)$,
$(-1,0)$
$u = 4$

Square, outward
coordinates $(-1,-1)$,
$(-1,1)$, $(1,1)$, $(1,-1)$,
$(-1,-1)$
$u = 4$

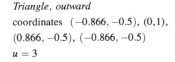

Triangle, outward
coordinates $(-0.5, 0.866)$, $(1,0)$,
$(-0.5, -0.866)$, $(-0.5, 0.866)$
$u = 3$

Triangle, outward
coordinates $(-0.866, -0.5)$, $(0,1)$,
$(0.866, -0.5)$, $(-0.866, -0.5)$
$u = 3$

Figure 3.12 Fractal island

a. Baseline:

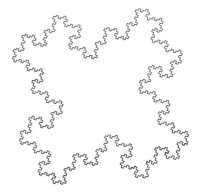

Square with vertices
(±1, ±1).
Motif:

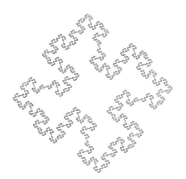

with intermediate points (0.4, 0.2),
(0.6, −0.2).
$u = 4$, $v = 3$, $p = 4$.

Figure 3.13 Minkowski island

b. Baseline as in *a*.
Motif:

with intermediate points (1/4, 1/4),
(3/4, −1/4).
$u = 4$, $v = 3$, $p = 5$.
The result, a variant of the
previous figure, again shows
Minkowski's meandering line.

Figure 3.14 Island of fjords

c. Baseline as in *a*.
Motif:

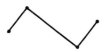

with intermediate points (0.3, 0.1),
(0.7, −0.3).
$u = 4$, $v = 3$, $p = 5$.
Even though the motif differs only
slightly from that of the previous
example, the result is very different.

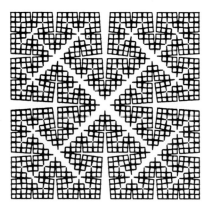

Figure 3.15 Torn square

d. Baseline as in *a*.
Motif:

with intermediate points (0.47, 0),
(0.5, 0.47), (0.53, 0).
$u = 4$, $v = 4$, $p = 4$.

Figure 3.16 Ice square

e. A variant on d. With the
same baseline but with
this motif:

with intermediate points (0.5, 0),
(0.5, 0.33), (0.5, 0).
$u = 4$, $v = 4$, $p = 5$.
The result reminds us of
ice crystals.

Figure 3.17 Ice triangle

f. Real ice crystals, snowflakes,
are built up on a hexagonal
pattern, with angles of 60°.
This results in the following
triangular version of e.
Baseline:

Motif:

with intermediate points (0.5, 0),
(0.375, 0.2165), (0.5, 0),
(0.625, 0.2165), (0.5, 0).
$u = 3$, $v = 6$, $p = 4$.

Figure 3.18 Construction of Lévy's fractal

g. The most simple motif is
the carpenter's try square.

Here each line-segment is
replaced again and again
by half a square. It is
fun to show a few initial
phases of the fractal's
construction:

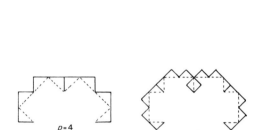

We see that the meandering
line forms a loop in the
fifth phase. This phenomenon
will become more prominent
later on.

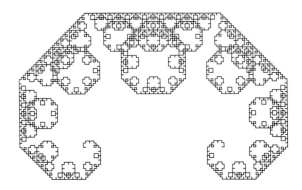

Figure 3.19 Lévy's fractal

Figure 3.19 shows us the 12th phase.
The data are
Baseline:

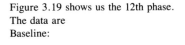

Motif:

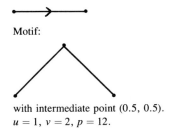

with intermediate point (0.5, 0.5).
$u = 1$, $v = 2$, $p = 12$.

Lévy's Curve

The curve of Figure 3.19—or, rather, its limit—is known as Lévy's
curve. The French mathematician Paul Lévy (1886–1971) was one of
the first scientists to do research on the figures that were later called
fractals. The interesting thing about the approximation of Figure 3.19,
the twelfth step, is that in it all previous approximations can be found.
For this we just have to look at the beginning of the meandering line
(through a magnifying glass) and successively count out 1, 2, 4, 8, 16,
32, ... line-segments of equal length.

And now for a small surprise. For base we select a square, so that
Lévy's curve is traced four times. The result (Figure 3.20) is a beautiful
symmetric figure in which we can hardly see how we started.

Figure 3.20 Lévy's tapestry

Lévy's curve can also be treated in the same manner as von Koch's
curve. The positions of the $2 \uparrow p$ line-segments of the pth approximation are determined by a binary expansion. The index n of a particular
line-segment, counting from the beginning, is written in binary:

$$n = t_0 + t_1 \cdot 2 + t_2 \cdot 2^2 + t_3 \cdot 2^3 + \cdots + t_{p-1} \cdot 2^{p-1}.$$

We then calculate the sum s of all binary digits:

$$s = t_0 + t_1 + t_2 + t_3 + \cdots + t_{p-1}.$$

The direction of the nth line-segment will then be $\phi = s\pi/2$ clockwise.

This may be enough for the trained mathematician. But since we take
it not everyone will understand the formulas, we will say the same thing
in a different way. For this, all that is needed is some knowledge of
the binary system.

For any number n we define s to be the number of ones in the binary
expansion of n. For instance: $n = 105$ (decimal) $= 1101001$ (binary)
$\rightarrow s = 4$.

The first few values of s can be read from the following table:

n	0	1	2	3	4	5	6	7	8	9	...
s	0	1	1	2	1	2	2	3	1	2	...
	→	↓	↓	←	↓	←	←	↑	↓	←	...

The value of s modulo 4 (the remainder after dividing s by 4) gives the direction of the line-segment. So:

$$s = 0, 4, 8, \ldots \quad \text{to the right,}$$
$$s = 1, 5, 9, \ldots \quad \text{downward,}$$
$$s = 2, 6, 10, \ldots \quad \text{to the left,}$$
$$s = 3, 7, 11, \ldots \quad \text{upward.}$$

Figure 3.21 shows us how this works. It differs from the previous method in that here we proceed from the initial point in a fixed direction, to the right, having selected the length of the steps beforehand. Without a computer, the meandering line can easily be traced on graph paper. We soon notice, though, that we are getting further and further away from the initial point. We could continue this way indefinitely— to infinity, in fact! Program LEVY in Appendix B works according to this method. It has the advantage of needing very little memory space.

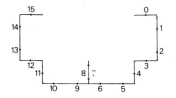

Figure 3.21 Lévy's fractal and the binary system

The Dragon Family

We fold a long, narrow strip of paper in two, so that we get a sharp crease [Figure 3.22(a) and (b)]. We repeat this twice, folding in the same direction. After unfolding, we have a strip consisting of eight

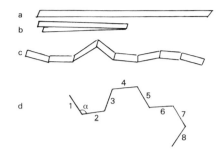

Figure 3.22 Folded strips of paper

pieces [Figure 3.22(c)]. Looking at it from the side, we see a meandering line, the vertices of which are the folds [Figure 3.22(d)].

In a situation like this, a mathematician will always wonder how it continues and what laws are at the bottom of it. One can continue with this experiment a bit longer, but not for very long. After ten steps the paper has already increased more than a thousand times in thickness $(2\uparrow10 = 1024)$. When we look at the meandering line of Figure 3.22(d), keeping in mind what we know about von Koch's and Lévy's curves, then the only thing that matters as far as the structure is concerned is which way we should continue after we have already drawn a sequence of 1, 2, ..., 8 line-segments. Should we turn right or left? We have kept the angle at each fold the same, and called it α. We can make the following list:

n	1	2	3	4	5	6	7	8	
d	1	1	−1	1	1	−1	−1		(1)

Here $d = 1$ means to the left, and $d = -1$ to the right. As yet no regularity can be detected. So we go on for a while:

n	...	9	10	11	12	13	14	15	16	
d	...	1	1	−1	−1	1	−1	−1		(1)

To discover the system, we have to realize the essential part factors of 2 play while we are folding. We then notice that $d(16) = d(8) = d(4) = d(2) = d(1) = 1$, that $d(12) = d(6) = d(3) = -1$, and that $d(10) = d(5) = 1$. This way we arrive at the following rule:

$$d(n) = 1 \qquad \text{for } n = 1, 5, 9, 13, \ldots$$
$$d(n) = -1 \qquad \text{for } n = 3, 7, 11, 15, \ldots$$
$$d(n) = d(n/2) \quad \text{for even } n.$$

DRAGON CURVE. Following this rule, we can draw the meandering line that results from folding repeatedly any number of times—a nice task for the computer. The program DRAGON is based on that rule. It produced the following illustrations.

First Figure 3.23: this shows the meandering line with the vertex angle $\alpha = 100°$ and $p = 6$. The figure then comprises $2 \uparrow 6 = 64$ line-segments.

Next, Figure 3.24, drawn with another version (DRAGON0) of the program for the special case $\alpha = 90°$. Here a rescaling procedure ensures that the meandering line always starts at $(0,0)$ and stops at $(1,0)$. For Figure 3.24 we have taken $p = 14$, so that $2 \uparrow 14 = 16384$ line-segments have to be drawn.

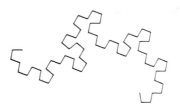

Figure 3.23 A piece of dragon line

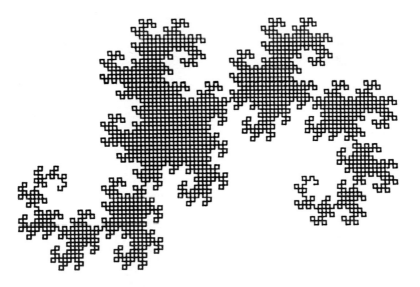

Figure 3.24 Dragon curve

The meandering line of Figure 3.24 reminded its discoverer, J. E. Heighway, of Chinese dragons; thus it is called the dragon curve. To our great surprise, we find that the meandering line does not intersect itself and also that it fills in the part of the plane occupied by the dragon regularly. Anyone who wants to do computer experiments has to bear in mind, as we mentioned before, that the resolution of the screen limits the choice of order. However, with $p = 10$ or $p = 12$ (preferably even numbers), for instance, we get good results.

It is nice to round off the right angles a little, as shown in Figure 3.25. This can be done with the program DRAGON1.

Figure 3.25 Rounded dragon curve

Figure 3.26 gives the result when $p = 10$.

Figure 3.26 Rounding the dragon curve

We can also get the dragon curve from the motif of Lévy's curve, the carpenter's try square. The only difference is that during the construction of each phase, each line-segment has to be replaced alternately by a try square and by its mirror image.

This principle of alternating the motif and its mirror image can of course also be applied to other motifs. Many variations are possible. We leave it to the ingenuity of the reader to design other beautiful meandering lines.

CHAPTER 4

Spirals, Trees, and Stars

This chapter deals with subjects the Greeks worked on two thousand years ago. Archimedes (287–212 B.C.) wrote a treatise on spirals; one kind of spiral bears his name even now. Pythagoras (569–500 B.C.) studied the famous figure in which squares are placed on the sides of a right triangle. He proved that the sum of the areas of the squares on two of the sides equals the area of the square on the hypotenuse. Actually, much earlier the Babylonians knew of this famous theorem, called the Pythagorean theorem ever since the time of the Greeks. A number of clay tablets with mathematical inscriptions in cuneiform bear witness to this. In our century this figure of Pythagoras has grown into a "tree" and spirals have become the building blocks of fractals.

Spirals are also the building blocks of the living world. The cell nucleus consists of a long, spirally wound structure, the nucleic acid or DNA, carrier of the genetic code, a building scheme for an organism yet to be formed. Organisms themselves can have a spiral structure too, like the ammonite in Figure 4.1, for example. This is a fossil dating from the Devonian Period, about three hundred million years ago. The same structure can still be found in countless species of shellfish alive now. In Figures 4.2 and 4.3 we show a few examples of these. These spirals are all associated with growth, with growing organisms. That's why this kind of spiral is called a *growth spiral*. In inanimate nature too we can find spiral structures; just think of a spiral galaxy, for instance.

The growth spiral embraces a microcosmos and a macrocosmos in its endless rotation. The Ashanti, a tribe on the Gold Coast in present-day Ghana, used the spiral motif on their gold weights, ornate copper or bronze castings for weighing gold dust. In Figure 4.4 we can see such a weight "charged with cosmic power." The growth spiral is self-similar and looks the same when magnified. That is why it can be seen as a kind of protofractal.

Figure 4.1 Ammonite

Spirals

There are three common types of planar spirals that have special names. Of these three, the most important by far for our purposes is the growth spiral. However, first we will say something about the other two, the evolute or unwinding spiral and the spiral of Archimedes.

EVOLUTE SPIRAL. When we unwind a fixed spool keeping the thread tight all the time, the end of the thread will trace a spiral. This is known as the evolute spiral or the *unwinding* of the circle. The evolute spiral of Figure 4.5 results when we run the program UNWIND.

Figure 4.2 Architectonica maxima

Like other programs for making spirals, UNWIND is based on the use of *polar coordinates* r and ϕ. At times these are a useful alternative to the Cartesian coordinates x and y. If P is a point fixed by coordinates (x, y) with respect to a Cartesian coordinate system XOY (Figure 4.6), then r equals OP, the distance from P to the origin O, and ϕ equals the angle between OP and the positive X-axis. So $r = OP$; $\phi =$ angle POX. We call r the radial distance and ϕ the polar angle.

The relationship between the two kinds of coordinates is, as we can see from Figure 4.6,

$$x = r \cos \phi, \qquad y = r \sin \phi.$$

Conversely, with the Pythagorean theorem:

$$r = \sqrt{x^2 + y^2}.$$

Figure 4.3 Shells and spirals

Figure 4.4 Ashanti gold weight

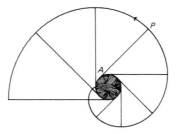

Figure 4.5 Unwinding of a circle

Note that r is always taken positive. The value of the polar angle is taken to increase counterclockwise from zero, starting on the positive X-axis. One complete rotation corresponds to an increase of ϕ by 2π radians.

Unfortunately the mathematical description of the evolute circle is a bit complicated. If in Figure 4.5, A, with polar coordinates (a, ϕ), is

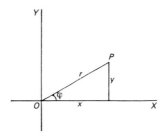

Figure 4.6 Polar coordinates

a point on the circle being unwound, then the location of the point P on the thread will be

$$x = a(\cos \phi + \phi \sin \phi),$$

$$y = a(\sin \phi - \phi \cos \phi).$$

We will not go into the proof of these formulas; after all, the computer just has to apply them. And it does not have any problems with that!

SPIRAL OF ARCHIMEDES. Perhaps the most simple spiral is the spiral of Archimedes (Figure 4.7; drawn with ARCHI). The grooves of

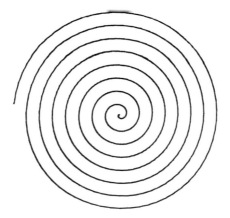

Figure 4.7 Spiral of Archimedes

a record and the windings of a coiled ribbon form this kind of spiral. Typically there is always a beginning, and the coils are everywhere equidistant from one another. That is why the spiral of Archimedes can be interpreted as the result of a revolving motion (uniform circular motion) combined with a constantly continuing linear motion (uniform linear motion). Imagine for instance an ant walking at constant speed in a straight line from the center of a record toward the edge. If we switch the record on while the ant is walking on it, then we, stationary observers, will see the ant move in a spiral of Archimedes. The mathematical description of this spiral in polar coordinates expresses this in the simple formula $r = a\phi$. After one revolution the polar angle ϕ will have increased by 2π, while the radial distance r has increased by $2\pi a$, which is precisely the constant distance between two coils.

All spirals of Archimedes are mutually similar; they differ only in scale. So in fact there is only *one* spiral of Archimedes. In Figure 4.8 we can see how a spiral of Archimedes can be used to derive a steady shuttling motion from a circular one. Two curve-segments of a spiral of Archimedes are mounted as mirror images on a revolving disk. Consequently the shaft that is driven by it will shuttle steadily. This is used in sewing and spinning machines, for instance.

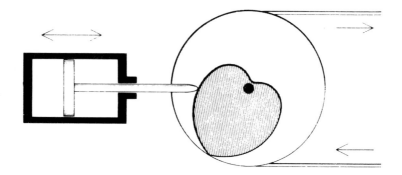

Figure 4.8 Converting a circular motion into a shuttling linear motion

GROWTH SPIRAL. For us the most important spiral is the growth spiral, so we will look into it more deeply. Another name for growth spiral is logarithmic spiral, owing to its mathematical description.

The spiral of Archimedes is defined by the formula $r = a\phi$. By replacing r by $\log r$, we get the formula for the growth spiral, $\log r = a\phi$.

Mathematics of the Growth Spiral

In Figure 4.9 three consecutive points, P_1, P_2, and P_3, of the spiral around O have been drawn. For convenience we have called their polar radii r_1, r_2, and r_3.

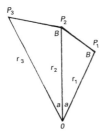

Figure 4.9 Three consecutive points of a logarithmic spiral

If for the moment we assume this to be a spiral of Archimedes, then r_1, r_2, and r_3 will form an arithmetic sequence. That means that the difference $r_2 - r_1$ equals the difference $r_3 - r_2$. Furthermore, the arithmetic mean of r_1 and r_3 then equals r_2, i.e., $r_2 = (r_1 + r_3)/2$.

The figure is drawn for a logarithmic spiral, however. This means that in the expression just given we have to replace r_1, r_2, and r_3 by their logarithms. Consequently $\log r_1$, $\log r_2$, $\log r_3$, ... will form an arithmetic sequence, so that $\log r_3 - \log r_2 = \log r_2 - \log r_1$, and so on. From the latter equation it follows that $\log(r_3/r_2) = \log(r_2/r_1)$ and so $r_3/r_2 = r_2/r_1$. This is the defining property of a geometric sequence. What is more, $r_2^2 = r_1 r_3$ or $r_2 = \sqrt{r_1 r_3}$, making r_2 the geometric mean of r_1 and r_3. To put it briefly, the polar radii of consecutive points of the logarithmic spiral P_1, P_2, P_3, ... with polar angles increasing by a constant quantity form a geometric sequence. The polar radii increase or decrease by a constant factor each time, depending on which direction we follow the spiral.

Looking at Figure 4.9 again, we can now conclude that $OP_1 : OP_2 = OP_2 : OP_3$. Basic geometric considerations now lead us to conclude that the triangles OP_1P_2 and OP_2P_3 are similar, and of course the same goes for all subsequent triangles. The triangle OP_2P_3 has originated from the triangle OP_1P_2 by a so-called similarity transformation, a combination of a rotation around the origin O and a change of scale. This could be either scaling up, as shown in Figure 4.9 ($r_2 > r_1$), or scaling down ($r_2 < r_1$). It is even possible that everything stays the same size ($r_2 = r_1$). In the first case the point-sequence P_1, P_2, P_3, ... will spiral toward infinity; in the second case it will converge to the origin O, but continue spiraling forever. In the third case the points will simply lie on a circle.

In the program LOGSPIRA we do not use the formula $\log r = a\phi$, but rather the equivalent $r = \exp(a\phi)$. The nature of the resulting spiral then depends on the sign of a:

$$a > 0 \quad \longrightarrow \quad \text{spiraling to infinity,}$$
$$a < 0 \quad \longrightarrow \quad \text{spiraling to the center,}$$
$$a = 0 \quad \longrightarrow \quad \text{circle.}$$

The growth factor depends on the change in the polar angle ϕ. If ϕ increases by α each time, then the polar radius will keep changing by the factor $\exp(a\alpha)$. We can interpret the point-sequence P_1, P_2, P_3 in Figure 4.9 and its continuation as the positions of a ship on the high seas sailing around a lighthouse at O. The captain fixes a course from P_1 by sailing in a direction that makes a fixed angle β with the line from ship to lighthouse. After sailing straight on for a while, the ship arrives at P_2. Meanwhile the angle between the line from ship to lighthouse and the course of navigation has increased from β to $\beta + \alpha$, so the course has to be adjusted. This repeats itself at P_3 and so on.

Whether the ship will approach the lighthouse or sail away from it depends on the values of α and β. In the borderline case the ship will sail around the lighthouse in a circle. In Figure 4.9 this would happen if the triangle OP_1P_2 were isosceles. In that case $\alpha + 2\beta = \pi$. From this it follows that for $\alpha + 2\beta < \pi$ the course will converge to the lighthouse, and for $\alpha + 2\beta > \pi$ it will diverge to infinity. The latter will of course happen only on a wet and flat planet of infinite size, but in mathematics, as in science fiction, much is possible.

Using the program LOGSPIRA we can draw such a sea route (Figure 4.10). Here the ship spirals toward the lighthouse. If we run the

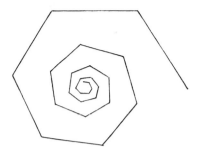

Figure 4.10 Broken logarithmic spiral

program, repeatedly reducing α, eventually the course will become so smooth that it cannot be distinguished from a true mathematical spiral (Figure 4.11).

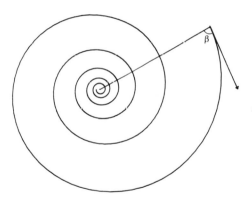

Figure 4.11 Logarithmic spiral or growth spiral

Finally the ship will sail a course that is a pure spiral. The angle between the direction of sailing and the line from ship to lighthouse does not change, and has a constant value β. The captain has to adjust his course continuously.

For a mathematical analysis, it is useful to know exactly how β depends on the coefficient a in the formula $r = r_0 \exp(a\phi)$. Comparing this with the previous formula, we see that a scale factor r_0 has been introduced. This implies that we take ϕ to be 0 when r has the value r_0. The formula for β is $\beta = \frac{1}{2}\pi + \arctan a$. From this it follows that for $\beta < \frac{1}{2}\pi$, corresponding to $a < 0$, the ship traces a spiral converging inward; and for $\beta > \frac{1}{2}\pi$, one diverging outward.

EADEM MUTATA RESURGO. The logarithmic spiral has always fascinated mathematicians and artists. The famous Swiss mathematician Jacob Bernoulli (1654–1705) studied it intensively and called it "spira mirabilis," the miraculous spiral.

Bernoulli thought the self-similarity, the fact that the spiral looks exactly the same on every scale, to be the greatest marvel of all. He noticed that an increase or decrease in the scale of the spiral is exactly the same as a rotation of the spiral as a whole. This property can be derived directly from its mathematical formulation. Rotation of the spiral through an angle γ means that we have to replace ϕ by $\phi - \gamma$. The rotated spiral is then given by

$$r = r_0 \exp[a(\phi - \gamma)] = (r_0 e^{-a\gamma}) e^{a\phi}.$$

The resulting formula differs from the original formula only in the scale factor $\exp(-a\gamma)$; that is, it is the same spiral on a different scale. Bernoulli thought this property so miraculous that he had the words "eadem mutata resurgo" (though transformed, I will rise again unchanged) inscribed on his tomb in the cathedral of Basel.

Spherical Spiral

Let us return for a moment to the idea of the ship sailing on a planet covered with water. This time we will be more realistic; we assume the planet is spherical like our own earth. Once again we have the ship following a fixed course in relation to a fixed point. For convenience we take this to be the North Pole. This enables us to use sea charts in a Mercator projection. With this projection, named after the famous Flemish cartographer Gerard Kremer (1512–1594)—"mercator" being Latin for kramer, merchant—all angles on the surface of the globe are represented in a plane, the chart, without distortion. Of course this method does have a drawback: regions near the poles are shown much larger than regions of equal area near the equator.

We can visualize the Mercator projection as the projection of a sphere from its center onto a vertical cylinder surrounding it. The poles end up at infinity. If we cut this cylinder lengthwise and unroll the surface area onto a plane, we get a vertical strip of infinite length: a Mercator projection of the entire surface of the sphere.

On this chart the route of a ship holding a fixed course (in other words, with the meridians being crossed at a constant angle not equal to a right angle) will look like a straight line. To counteract the drawback of the map's stopping at the same meridian both on the right and on the left, we repeatedly place a copy of the vertical strip on each side. This way, in theory at least, the result will be an infinite series of identical Mercator projections, each one continuing into the next. With this arrangement of charts, a course of navigation along a sloping straight line can be represented in its entirety. When we follow this course on the surface of the sphere, a spiral curve, a so-called spherical spiral or loxodrome, will result. The spherical spiral runs from pole to pole in an infinite number of windings that get smaller and smaller near the poles. Figure 4.12 shows us a spherical spiral made by the program SPHERSPI.

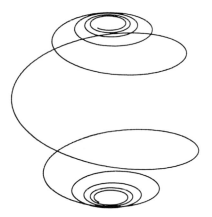

Figure 4.12 Spherical spiral

Spherical spirals also arise with two ships (say, a merchant and a pirate), if the pirate is sailing in the direction of the merchant and the merchant is trying to avoid the pirate, holding a fixed course. The ships will spiral around one another, the distance between them either increasing or decreasing.

We now examine a similar situation in a plane more thoroughly. We imagine four dogs A, B, C, and D running after each other at equal speed. Initially they are situated at the vertices of a square. A runs toward B, B toward C, C toward D, while D runs after A. The dogs finish up running in logarithmic spirals that meet in the center of the square after having turned about one another an infinite number of times.

Our picture of this (Figure 4.13) was drawn in a slightly different way. There the square is repeatedly subjected to the same similarity transformation. It has been rotated and reduced each time in such a way that the vertices of the new square lie on the sides of the previous one. The computer program for Figure 4.13 can be found under WHIRL.

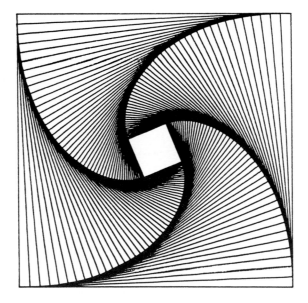

Figure 4.13 Square whirlpool

The Pythagoras Tree

In 1957 a remarkable book entitled *Het wondere onderzoekingsveld der vlakke meetkunde* (*Geometry in the Plane: A Miraculous Field of Research*) appeared. In it the author, A. E. Bosman (1891–1961), tried to convey to young people his wonder at the miraculous geometrical shapes of nature. In this work the spiral motif figures prominently. Even the cover shows the spiral of a nautilus shell. As a textbook it was not a success—the work reads more like an essay. If Bosman were alive and working now, he might have made important contributions to computer art. One of the most remarkable figures Bosman made is the Pythagoras tree, drawn by him during the Second World War on the same drawing board he used for designing submarines. Now, a few decades later, we can make his design appear on the screen within minutes. Figure 4.14 shows us a photograph of it.

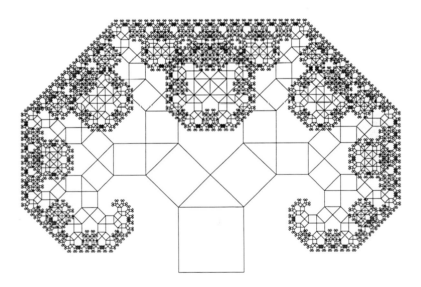

Figure 4.14 Pythagoras tree

In Chapter 1 we saw that the construction of such a tree is based on the binary number system. If we number the squares as in Figure 4.15, we find that a square with a given index n supports an isosceles right triangle from which two smaller squares sprout. The one on the left has index $2n$, an even number; the one on the right, $2n + 1$, an odd number. Together these three form a geometrical representation of the Pythagorean theorem, which states that the smaller squares together have the same area as the larger square. As the smaller squares are equal here, each one is half the size of their shared predecessor. If the area of the original square is 1, then the total area of squares 2 and 3 will also be 1. Going on, the same holds for each smaller series. So in Figure 4.15, the eight squares 8, 9, 10, ..., 15 together have the same area as the original base square.

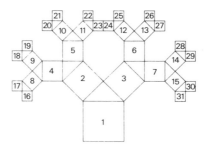

Figure 4.15 First stages of the Pythagoras tree

The position of an arbitrarily chosen square with index n depends on the divisibility of n by 2. We find that number 13 takes a right-hand position in relation to number 6, number 6 a left-hand one in relation to number 3, and number 3 a right-hand one in relation to number 1. We can obtain this from the binary notation of 13: $13 = 1101$. Reading from left to right and omitting the left-most one we get 1 (right), 0 (left), 1 (right). In the program based on this, PYTHT1, the squares are drawn in the order 1; 2, 3; 4, 5, 6, 7; 8, 9, ..., 15 and so on. The square with index n is generated from the original square by a sequence of similarity transformations according to the binary digits of n. Since this program includes a lot of arithmetic it is a bit slow, especially if we want to go up to the order of 10, say. In that case $(2 \uparrow 11) - 1$ squares have to be drawn!

There are better methods for achieving this. Our working rule is that economy in calculations will be achieved at the expense of memory space. In the program PYTHT2 this economy is obtained by storing in memory all kinds of intermediate results for later use. The order of the tree that can be attained is then determined entirely by the available memory space.

In PYTHT3 the tree is built up by using the backtrack method, a clever and most efficient method that will be discussed in the next chapter. The programs constructed with this method are quick and need little memory space.

LOPSIDED PYTHAGORAS TREE. The Pythagoras tree can be varied and extended in many ways. An obvious generalization, which Bosman found, is a lopsided tree. Here the square is followed by an arbitrary right triangle. The initial three are sketched in Figure 4.16. The curled shape is formed by continuing each time with the square on the right. The curl is a genuine logarithmic spiral, determined by a repeated similarity transformation. In Figure 4.16 the center C of the spiral has been marked. That similarity transformation, **R**, is a rotation through an angle $-\alpha$, the angle in the right triangle, combined with a reduction in scale of $\cos \alpha$. On the left-hand side, not drawn in Figure 4.16, another similarity transformation, **L**, operates analogously. This is a rotation through the angle $\pi/2 - \alpha$ combined with a reduction of $\sin \alpha$.

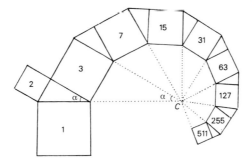

Figure 4.16 Logarithmic spiral in a lopsided Pythagoras tree

The Pythagoras tree is a fine example of a mathematical fractal, a self-similar figure in which each square can be interpreted in its turn as the trunk of an entire tree. We can specify this self-similarity here by saying that both the rotation to the left, **L**, and the rotation to the right, **R**, carry the tree into itself. The only effect this has is a kind of shifting of the squares, and this we notice at the trunk. At the tips, the smallest squares, we do not notice anything. The squares with indices 1, 2, 4, 8, 16, . . . , i.e., the powers of 2, form a spiral converging to the center of **L**. If we subject this figure to the transformation **R**, a second spiral results: the one composed of the squares 3, 6, 12, 24, 48, . . . This second spiral is part of the tree too. We can do this once more, or better still, we can subject the original spiral to an arbitrary combination of **L** and **R** transformations. Each combination produces another spiral as part of the tree. Thus we find that the Pythagoras tree is full of spirals; infinitely many of them. Detecting them in pictures of the trees is often difficult, because in the later stages of formation the spiraling squares cover one another more and more. In Figure 4.17 a lopsided tree has been drawn with the program PYTHT3, taking $\alpha = \pi/6$ (30°).

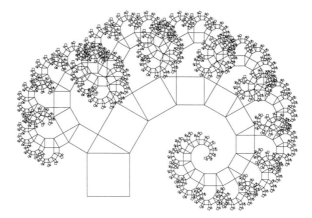

Figure 4.17 Lopsided Pythagoras tree

MORE VARIATIONS. Bosman also hit upon the idea of following the squares alternately by a scalene triangle and its mirror image. Figure 4.18 shows us the result.

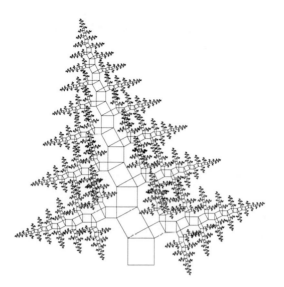

Figure 4.18 Alternating Pythagoras tree

We can simplify the Pythagoras tree by omitting the squares and just drawing line-segments that join the centers of the triangles. The triangles themselves disappear. This way we get a tree that is rather bare (Figure 4.19). For this tree we can use almost the same computer program as for the ordinary tree. It can be found in Appendix B as PYTHB.

In Mandelbrot's book there are other variations of the Pythagoras tree; see Figure 4.21, for instance. This version is based on the trunk shape of Figure 4.20. The side branches, similar copies of the trunk, are attached between the points *A* and *B* and the points *E* and *F*. The program for this, TREEM, is a variation of PYTHT3. In it the branches are attached both ordinarily *and* mirrored each time. Also, the parameters have been chosen in such a way that in the limit the figure will fill the plane.

We finish by showing a more realistic fractal tree (Figure 4.23). As in Mandelbrot's tree the construction is based on a model of a trunk with points of attachment for the side branches (Figure 4.22).

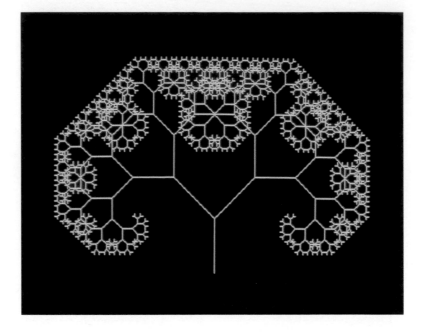

Figure 4.19 Branching Pythagoras tree

Figure 4.20 Trunk of the Mandelbrot tree

Stars

Figure 4.24 shows a star fractal. This consists of a regular star pentagon, a pentagram, with a garland of five smaller specimens. Each of these five smaller pentagrams supports on its four loose points still

Figure 4.21 Mandelbrot tree

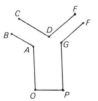

Figure 4.22 Model for a realistic tree

smaller pentagrams. Theoretically this can continue indefinitely. The resulting star fractal again has the defining property of self-similarity. In it we can also find broken spirals. In actual practice, of course, we draw with limited accuracy and have to stop after a few steps. Figure 4.24 shows us five phases; with its four successive reductions it already consists of 1280 line-segments!

Figure 4.23 Realistic tree

This star fractal is built up as one closed line, successive line-segments always meeting at the same angle α. A fragment with $\alpha = 4\pi/5$ (144°) is sketched in Figure 4.25. Suppose the line-segments of Figure 4.24 are numbered from $n = 0$ to $n = 1279$. If the first line-segment, with index $n = 0$, has the direction $\phi = 0$, then the direction of a line-segment with an arbitrary index n will be $n\alpha$. The only thing that matters is the rule that tells us the length of the nth line-segment. In this case there are five different lengths: 1, r, r^2, r^3, r^4, in which r is a reduction factor (in Figure 4.24 equal to 0.35). The rule on which Figure 4.24 is based is

$$n = 1, 2, 3, 5, 6, 7, 9, 10, 11, 13, \ldots \qquad \text{length } r^4,$$
$$n = 4, 8, 12, 20, 24, 28, 36, 40, 44, 52, \ldots \quad \text{length } r^3,$$
$$n = 16, 32, 48, 80, 96, 112, \ldots \qquad \text{length } r^2,$$
$$n = 64, 128, 192, 320, 384, 448, \ldots \qquad \text{length } r,$$
$$n = 0, 256, 512, 768, 1024, \ldots \qquad \text{length } 1.$$

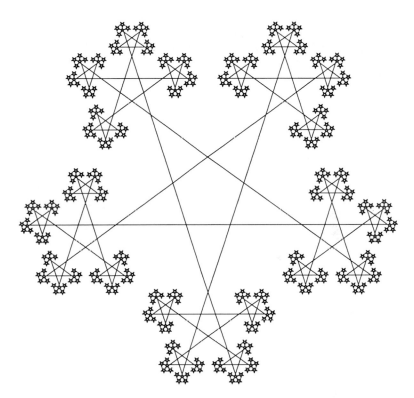

Figure 4.24 Star fractal

From this we can see that the length of the line-segment with index
n depends on the number of factors of 4 in n. We can quite simply
generalize this rule so that we can draw other star fractals. We call the
number of phases p, and we take an arbitrary number ν instead of 4.
This gives us the general rule

n has no factor ν length r^{p-1},
n has one factor ν length r^{p-2},
n has two factors ν length r^{p-3},
n has three . . .
. . .
n has at least $p-1$ factors ν length 1.

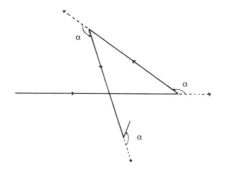

Figure 4.25 Scheme of the star fractal

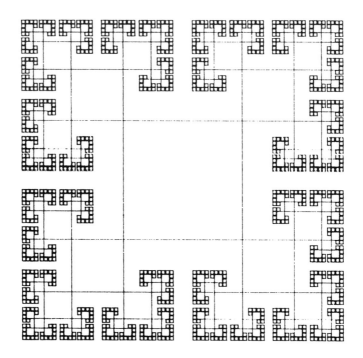

Figure 4.26 Square star fractal

In the general situation we count from 0 to $(\nu+1)\nu \uparrow (p-1)$. There are exactly $\nu+1$ line-segments of the biggest length, 1; $(\nu+1)(\nu-1)$ of length r; $(\nu+1)(\nu^2-\nu)$ of length r^2; and so on. So Figure 4.24 was constructed with $p = 5$, $\nu = 4$, $\alpha = 4\pi/5$, and $r = 0.35$. With $p = 7$, $\nu = 3$, $\alpha = \pi/2$, and $r = 0.47$, we get an entirely different picture (Figure 4.26). Both figures are made with the same computer program, STAR. In it, arbitrary values can be chosen for p, ν, α, and r.

The Analysis of a Fractal

Fractals are geometrical figures with a built-in self-similarity. In this chapter we will explain exactly what this means. We will find that fractals can be defined as point-sets that are invariant under a semigroup of contractions. This is a mathematical definition that needs explanation. By invariance with respect to a transformation we mean that if P is a point of the fractal, the point obtained by transforming P will also belong to the fractal. In the most simple case the contraction is a scale reduction (central enlargement), with or without a rotation. We have already met such similarity transformations in the chapter about spirals. Here we examine them once more, but this time more generally.

We denote a transformation by a capital letter, **L** or **R** for instance. A combination of two transformations **L** and **R** is itself a transformation. It is written as a product, to be read from right to left, **RL** (first **L**, then **R**) or **LR** (first **R**, then **L**), for example. This differs from ordinary products of numbers in that the order matters. Repetition of the same transformation is expressed as a power, **L** ↑ 2 for **LL** and **R** ↑ 3 for **RRR**. It is convenient to reserve the letter **I** for the identity transformation. This is the transformation under which everything stays in its place.

In mathematics, the collection of all the transformations (including the transformation **I**) that result by combining **L** and **R** in all possible ways is called a semigroup. Anticipating what will be said on the subject later in this chapter, we now give an example that in its simplicity is nevertheless representative of a much more general situation.

We examine once again the Cantor set we discussed exhaustively in Chapter 2. That set consists of discrete points that can be written in ternary as infinite fractions with zeros and twos only. For instance:
$x = 0.020\,220\,222\,02\ldots$

We now imagine every point x of the line-segment $[0, 1]$ subjected to two contractions:

$$\mathbf{L} : x \rightarrow \tfrac{x}{3} \qquad \text{so } [0, 1] \rightarrow [0, \tfrac{1}{3}]$$

$$\mathbf{R} : x \rightarrow \tfrac{2}{3} + \tfrac{x}{3} \quad \text{so } [0, 1] \rightarrow [\tfrac{2}{3}, 1]$$

In words: $\mathbf{L}$ transforms a point x into $x/3$, while $\mathbf{R}$ transforms a point x into $2/3 + x/3$. This means $\mathbf{L}$ is a central enlargement with center $x = 0$ and scale factor $1/3$. Because of this the interval $[0, 1]$ is transformed into the interval $[0, 1/3]$: it is shortened. $\mathbf{R}$ differs from $\mathbf{L}$ only in its center of enlargement; this is 1. In Figure 5.1 this is represented geometrically.

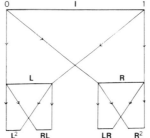

Figure 5.1 Diagram of combined transformations

If we write x as $0.c_1 c_2 c_3 c_4 \ldots$ in ternary, then

$$\mathbf{L}x = 0.0c_1 c_2 c_3 c_4 \ldots$$

$$\mathbf{R}x = 0.2c_1 c_2 c_3 c_4 \ldots$$

The string of digits shifts one place to the right and is supplemented by a 0 or a 2 on the left. If x belongs to the Cantor point-set, then $\mathbf{L}x$ and $\mathbf{R}x$ will also belong to it. Consequently, the complete set is mapped by $\mathbf{L}$ onto a part of itself. The same goes for $\mathbf{R}$ and also for all combinations of $\mathbf{L}$ and $\mathbf{R}$, the semigroup generated by $\mathbf{L}$ and $\mathbf{R}$. For this property we use the term "invariant," i.e., unchanging. Conversely, we can say that the Cantor set is determined precisely by this property of being invariant under two contractions and all combinations generated by them.

This principle will turn out to be extremely fruitful for the concept of "fractal." In this chapter we will analyze the Pythagoras tree, the Lévy curve, and the von Koch curve in this way. We start by discussing the concept of "dimension," the so-called "capacity." Here too we go into mathematical detail in greater depth.

The Dimension of a Fractal

When Mandelbrot introduced the concept of "fractal" in 1977, he also introduced the term fractal dimension, a concept of dimension he based on a definition given by Hausdorff in 1919. This concept of dimension as used by Mandelbrot is a simplification of Hausdorff's and corresponds exactly with the 1958 definition of Kolmogorov (1903–1987) of the "capacity" of a geometrical figure. As the concept dimension has different meanings in the literature—mathematicians use at least four other definitions—we will stick to Kolmogorov's terminology and use the word "capacity."

First we define the capacity of a point-set on a straight line, the number line, keeping in mind the example of the Cantor point-set. Kolmogorov thought of measuring such a point-set with an ever decreasing measuring unit h, just like we did when we discussed the coastline in Chapter 3. We imagine the points covered as sparingly as possible with line-segments of length h. Let us call the smallest number of line-segments needed for this $N(h)$. This is a function of h. Consequently, at a subsequent measuring, when we use a smaller h, N will change as well and will of course increase. The capacity is then defined as

$$\text{(A)} \qquad D = \lim_{h \to 0} \frac{\log N(h)}{\log\left(\frac{1}{h}\right)}$$

As we see, this definition corresponds to the similar one in Chapter 3. But now the definition concerns points on a straight line, whereas in Chapter 3 we were talking about a meandering line in the plane.

In order to apply the definition to the Cantor point-set, we take another look at the different stages of the construction (Figure 5.2). We start with a line-segment of length 1. We can cover this best (i.e., measure it) by taking exactly one line-segment also of length 1. In

Figure 5.2 Determining the fractal dimension of the Cantor fractal

the next phase, to be treated like the first, the line-segment is covered by two line-segments of length 1/3. Next $h = 1/9$ with $N = 4$, then $h = 1/27$ or $3 \uparrow (-3)$ and $N = 8$ or $2 \uparrow 3$. In the nth phase this will be $h = 3 \uparrow (-n)$ and $N = 2 \uparrow n$. Plugging this into formula (A) we get

$$D = \frac{\log 2}{\log 3} = 0.6309$$

In fact, we are still only repeating the calculation we made in Chapter 3. The analogous calculation in the plane, however, has something new to offer. As an example let us take the Minkowski sausage (Figure 3.9). We cover it with small squares of side h. We put the fractal on graph paper, as it were, and count the number of squares containing vertices of the fractal (Figure 5.3).

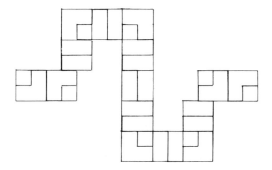

Figure 5.3 Covering the Minkowski fractal with squares

If we again call the smallest number of squares of size $h \times h$ needed to cover the fractal $N(h)$, then formula (A) applies. In this case $N(h)$ can be estimated quite easily.

To do this we examine the nth phase in the construction of the sausage. The original line-segment of length 1 has then been replaced by a meandering line consisting of $8 \uparrow n$ line-segments of length $4 \uparrow (-n)$. In Figure 5.4 we take a closer look at such a line-segment PQ.

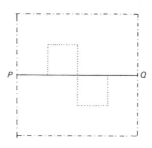

Figure 5.4 Detail of covering

Everything that happens within it in the subsequent phases—a copy of the fractal in miniature—can be covered by the square shown, the side of which is $h = 4 \uparrow (-n)$. There are $8 \uparrow n$ of these line-segments, so $N(h) = 8 \uparrow n$. Formula (A) immediately gives $D = \log 8 / \log 4 = 3/2$. The capacity of the Minkowski sausage is therefore 1.5.

In general we can use formula (A) for a fractal in the plane, provided we interpret $N(h)$ as the minimum number of squares of side h needed to cover the fractal completely.

The dimension, or rather the capacity, of a fractal does not say much about its outward appearance. For a smooth line such as a circle, both the ordinary dimension and the capacity will be 1. A fractal with $0 < D < 1$ certainly does not include line-segments and will therefore be discrete, just like the Cantor fractal. One could call a fractal like that pointlike or dustlike.

Instead of starting with a line-segment, we can start with a square and build a Cantor set in the same way. If we use coordinate geometry and take (0,0), (1,0), (0,1), and (1,1) as the vertices of the square, an arbitrary point of the "Cantor square" will correspond to coordinates (x, y), where both x and y belong to the classic Cantor set. Thus both

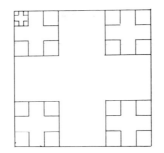

Figure 5.5 Square analogue of the Cantor fractal

the horizontal and the vertical projections of the fractal points in the square form an ordinary Cantor set. We can also do this construction geometrically (Figure 5.5) by repeatedly deleting the cross-shaped middle part. At the nth phase, $4 \uparrow n$ squares with sides $3 \uparrow (-n)$ remain. From formula (A) it then follows that $D = \log 4/\log 3 = 1.2619$.

We could describe a fractal with $1 < D < 2$ as a bold line, or a line that more or less fills the plane. In Chapter 3 we looked into the degree of meandering of a meandering line. For a line like the von Koch curve or the Minkowski meander, that definition of degree of meandering corresponds to the new definition of capacity. Problems will arise if the fractal is the final stage of an infinite series of approximations, and if those approximations are broken lines, say, that repeatedly intersect themselves. The Lévy curve (Figure 3.19) and the Pythagoras tree (Figure 4.14) are fractals of this type. A careful analysis shows that the capacity of the Lévy curve is exactly 1. By Mandelbrot's original definition it would not be a fractal. And that would be a pity. Let it be said again: the essential property of a fractal is indefinitely continuing self-similarity. Fractal dimension is just a by-product.

Similarity Transformations

We have already had a few opportunities to point out that built-in self-similarity is a defining property of fractals. When discussing the spirals of Chapter 4, we especially emphasized the significance of rotations and changes in scale. Here we will go into this more systematically.

Let us look at transformations of points P, Q, etc. in the plane; we shall refer to them by letters like **T**, **L**, **R**, and so on. By the notation **T**(P) we indicate that the transformation **T** is applied to the point P. We often indicate the transformed point by a prime accent: $P' = $ **T**(P).

Two transformations **T**$_1$ and **T**$_2$, carried out one after the other, result in a combined transformation, to be written as **T**$_2$**T**$_1$. This must be read from right to left: first **T**$_1$, then **T**$_2$. The order matters here, as **T**$_1$**T**$_2$ is often different from **T**$_2$**T**$_1$. This is called multiplication of two transformations, and for this the same notation is used as for multiplying two ordinary numbers. So if we multiply **T** by **T**, repeating the same transformation, it is not written as **TT**, but as **T** $\uparrow$ 2. Analogously **T** $\uparrow$ m is the notation for the transformation that results when **T** is repeated m times. The identity, the transformation that leaves everything in its place, is indicated by **I**. For instance, if geometrically speaking **T** is reflection in a given line, then **T** $\uparrow$ 2 $=$ **I** expresses in a simple way that the original situation is restored by another mirroring.

We will now discuss the main similarity transformations one by one. These are rotation, rescaling or central enlargement, rotation-enlargement, and reflection or mirroring. As far as the mathematical discussion is concerned, we restrict ourselves to simple geometrical principles and the use of coordinate geometry, as this is of vital importance in making computer programs.

Rotation

A rotation is determined by a center and an angle of rotation. The rotation angle is always measured in radians or degrees, taking the direction of rotation positive if it is counterclockwise. (Remember the fact that π radians equals $180°$.)

In coordinate geometry a rotation through an angle α with the origin as center (Figure 5.6) is expressed by

$$x' = x \cos \alpha - y \sin \alpha$$
$$y' = x \sin \alpha + y \cos \alpha$$

Here x and y are the Cartesian coordinates of P with respect to the XOY-system, and x' and y' those of the transformed point P'. The angle α is called the angle of rotation.

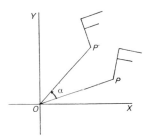

Figure 5.6 Rotation

Making a small change, we can also describe a rotation in which the point (x_0, y_0) is the center of rotation:

$$x' = (x - x_0)\cos\alpha - (y - y_0)\sin\alpha + x_0$$
$$y' = (x - x_0)\sin\alpha + (y - y_0)\cos\alpha + y_0$$

To test whether this formula is correct, we replace x and y by x_0 and y_0, respectively, on the right-hand sides. We then get $x' = x_0$ and $y' = y_0$. This means that under this rotation the point (x_0, y_0) remains where it is and is itself the center of rotation.

We can illustrate this by a look at two rotations, **L** and **R**. **L** is a rotation around A, $(0,0)$, with angle of rotation $\pi/6$ ($30°$), and **R** is a rotation around B, $(1,0)$, with angle of rotation $\pi/3$ ($60°$). In Figure 5.7 we can see that the rotation **L** (left) transforms the line-segment AB into AB_1. If we then let **R** (right) follow, AB_1 will be transformed into $A_2 B_2$. So:

$$\mathbf{RL}(AB) = A_2 B_2.$$

If we do the same swapping **L** and **R**, then $\mathbf{R}(AB) = A_2 B$ and

$$\mathbf{LR}(AB) = A_1 B_1.$$

This alone shows us the importance of the order when one is composing rotations with different centers. It almost goes without saying that the multiplication of two rotations generally is another rotation. Its rotation angle equals the sum of the angles of rotation of the components, so the result is a translation only if this sum is zero or 2π.

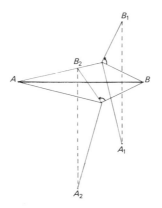

Figure 5.7 Two rotations combined

Determining the position of the center of rotation is a little more complicated, and we do not go into that here.

Note that in Figure 5.7 A_1B_1 and A_2B_2 are parallel and equally long, and that their position relative to AB has been rotated through $\pi/6 + \pi/3 = \pi/2$ (90°). For the sake of completeness, the rotation centers of **RL** and **LR** have also been indicated in Figure 5.7.

What we find here geometrically we can also derive with coordinate geometry. **L** and **R** are defined by

$$\mathbf{L} \begin{cases} x' = (\sqrt{3}x - y)/2 \\ y' = (x + \sqrt{3}y)/2 \end{cases}$$

$$\mathbf{R} \begin{cases} x' = (x - \sqrt{3}y + 1)/2 \\ y' = (\sqrt{3}x + y - \sqrt{3})/2 \end{cases}$$

From this it follows that

$$\mathbf{RL} \begin{cases} x' = -y + \dfrac{1}{2} \\ y' = x - \dfrac{\sqrt{3}}{2} \end{cases}$$

and

$$\mathbf{LR} \begin{cases} x' = -y + \dfrac{\sqrt{3}}{2} \\ y' = x - \dfrac{1}{2} \end{cases}$$

This tells us that both **RL** and **LR** are rotations of $\pi/2$ (90°), but that the locations of their rotation centers are different. A calculation gives us the positions $\left((1 + \sqrt{3})/4, (1 - \sqrt{3})/4\right)$ for **RL** and $\left((1 + \sqrt{3})/4, (-1 + \sqrt{3})/4\right)$ for **LR**.

Change of Scale

A rescaling or central enlargement is characterized by a center and a scale factor c. If, as in Figure 5.8, the origin O is the center and the point P is transformed into P', then P' will lie on the line OP. Also $OP' = c \cdot OP$.

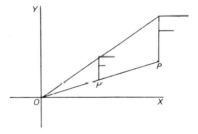

Figure 5.8 Central enlargement or rescaling

In fractals we usually have $|c| < 1$, so the transformation is a contraction. Thus a figure at P will be reduced to a figure at P'.

As a rule c is positive, though mathematically speaking there is nothing against taking c negative. In that case P' and P lie on either side of the center O. A rather exceptional case is $c = -1$. Here P' and P take mirrored positions in relation to O. The term for this is *point reflection*. This rather unusual transformation equals a rotation around O of π radians (180°).

In coordinate geometry a central enlargement with respect to the origin is expressed as

$$x' = cx \quad \text{and} \quad y' = cy.$$

If we want to describe a central enlargement with the point (x_0, y_0) as center, we use the formulas

$$x' = c(x - x_0) + x_0$$
$$y' = c(y - y_0) + y_0.$$

Rotation-enlargement

The most important similarity transformation is the rotation-enlargement, a rotation and a central enlargement combined. As we already discussed this transformation exhaustively in Chapter 4 when we were dealing with the logarithmic spiral, we can be brief here.

Here too the determining characteristics are a center and a scale factor always taken positive. The rotation-enlargement can always be regarded as the product of a rotation $\mathbf{R}$ and a central enlargement $\mathbf{C}$ with the same center. The order in which that product is taken does not matter: $\mathbf{RC}$ or $\mathbf{CR}$ (Figure 5.9).

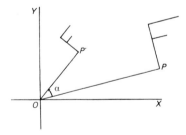

Figure 5.9 Rotation-enlargement

While analyzing a fractal we may come across the following geometrical problem. We have two similar figures, the triangles PQR and $P'Q'R'$, say (Figure 5.10). The question is, by what rotation-enlargement can the triangle PQR be transformed into $P'Q'R'$? We

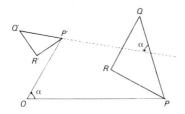

Figure 5.10 Center of rotation of two similar triangles

can easily find the angle of rotation α. For this we have only to look at two corresponding lines, PQ and $P'Q'$, for instance. Determining the center O is more complicated, but more interesting too. Here, however, we will restrict ourselves to remarking that the position of O can be derived from the fact that the angle $P'OP$ equals α and that $OP':OP = Q'P':QP = c$.

In coordinate geometry a rotation-enlargement with the origin O as center is expressed by

$$x' = ax - by$$
$$y' = bx + ay.$$

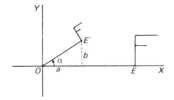

Figure 5.11 Rotation-enlargement in coordinates

The significance of the coefficients a and b becomes clear when we look at Figure 5.11. Here E is the point $(1, 0)$. From the formulas we see that this point is transformed into E' with coordinates (a, b). The scale factor of the transformation is therefore $OE'/OE = \sqrt{a^2 + b^2}$. The angle of rotation α satisfies $\cos \alpha = OA/OE' = a/\sqrt{a^2 + b^2}$ and $\sin \alpha = AE'/OA = b/\sqrt{a^2 + b^2}$.

It takes only one more step to get the formulas for the general rotation-enlargement with an arbitrary center:

(B)
$$x' = ax - by + c$$
$$y' = bx + ay + d.$$

Here the scale factor is $\sqrt{a^2 + b^2}$, and the angle of rotation α is determined by $\cos \alpha = a/\sqrt{a^2 + b^2}$ or $\sin \alpha = b\sqrt{a^2 + b^2}$.

These formulas are extremely important. They are found in many computer programs for making fractals.

APPLYING ROTATION-ENLARGEMENT. We now apply all this to the Lévy curve of Chapter 3 (Figure 3.19), which when continued indefinitely is the limit figure of the Pythagoras tree (Figure 4.14). In Figure 5.12 we sketched the first phases of its construction. In every subsequent phase, the line-segments of the broken line are replaced by try squares. The vertices stay where they are; once formed, they are a permanent part of the fractal. In fact, the fractal is composed of all these vertices.

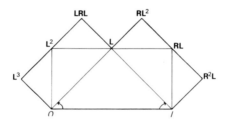

Figure 5.12 Similarity transformations in the Lévy fractal

In Figure 5.12 two rotation-enlargements can be detected. The first one, **L**, has O as center, $\pi/4$ as rotation angle, and $1/\sqrt{2}$ as reduction factor. The second one, **R**, has center I, rotation angle $-\pi/4$, and reduction factor $1/\sqrt{2}$ also. Both **L** and **R** transform vertices into other vertices. The situation is analogous to that of the Cantor point-set discussed in the introduction to this chapter. So the Lévy fractal is invariant under both **L** and **R**, and everything that can be derived from them by forming products.

The points of a fractal can be generated systematically by subjecting, for instance, the right-hand base point I to an arbitrary series of L and R transformations. For this purpose we can use the binary number system again by interpreting L as the binary digit 1 and R as the digit 0. This way the fractal consists of the same number of points as there are numbers between 0 and 1 in binary. As we know by now, there are infinitely many of these! The corresponding description in coordinates is

$$L \begin{cases} x' = (x - y)/2 \\ y' = (x + y)/2 \end{cases}$$

$$R \begin{cases} x' = (x + y + 1)/2 \\ y' = (-x + y + 1)/2 \end{cases}$$

So the center of L is $(0,0)$ and the center of R is $(1,0)$.

If we choose an arbitrary point P as starting point, and if we let the transformations L and R operate on it repeatedly, then twice as many points will be added at each step. We can picture this process as a tree with index numbers (Figure 5.13). If P is a point of the fractal, then all subsequent points derived from it will be points of the fractal

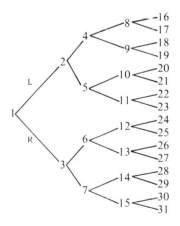

Figure 5.13 Numbering of a binary tree

too. This is one way to write a computer program for constructing fractals point by point. The method is fast and efficient, but it has the technical drawback of requiring a lot of memory space. For example, to calculate $2 \uparrow 14 = 16384$ points, in order to draw them on the screen or with the plotter, $2 \times 2 \uparrow 12 = 8192$ memory locations are needed, which is rather a lot.

There is a way to do this that requires much less memory space. For this we just have to number the tree differently (Figure 5.14). This requires only $4 \times 12 = 48$ memory locations. The method is called "backtracking"; it is incorporated in the program DUSTB. This program is exceptionally flexible, as it allows different choices of **L** and **R**. It is one of the most popular programs for making fractals according to a binary tree structure.

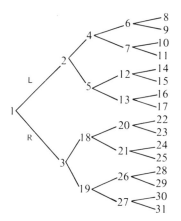

Figure 5.14 Numbering according to the backtrack method

With this method of enumeration we keep returning to the smallest branches and then backtrack. The latter is done in such a way that optimal use is made of partial results calculated earlier. We will now demonstrate how DUSTB is built up by looking at Figure 5.14.

The order chosen, p, is 4. In the figure this produces $p + 1$ levels numbered $s = 0$ to $s = p$. The indices then run from 1 to $2 \uparrow (p+1) - 1$. The program has a group-counter m running from 0 to $2 \uparrow (p-1) - 1$; it is 7 in this case. Starting from the initial position 1 at $m = 0$, the pairs 2, 3; 4, 5; 6, 7; 8, 9 are calculated, and the corresponding points are drawn. So the group consists of 4×2 points. The next group with $m = 1$ consists of the single pair 10, 11; it has 1×2 points. For $m = 2$, starting from the point 5 calculated earlier, the positions 12, 13; 14, 15; 16, 17 are determined, a group of 3×2 points. It goes on like this.

So we keep backtracking to a previous level. There a point calculated earlier will serve as the starting point of a new binary series. This shuttling back and forth is illustrated by the following table:

m	0	1	2	3	4	5	6	7
s	1	4	3	4	2	4	3	4
group	1–9	10–11	12–15	16–17	18–23	24–25	26–29	30–31

Apart from $s = 1$ for $m = 0$, we establish that s always equals p minus the number of factors of 2 in m. This observation, simple though it may seem, is not obvious. In a way, it is the secret behind the program. The reader would do well to work this out himself for $p = 5$.

The principle works just as effectively for trees with a ternary or higher-order structure. If each branch splits into ν smaller branches, then the heart of the program can be written as

for $m = 1$ to $2 \uparrow (p - 1) - 1 : s = p ; n = m$
if $n \bmod \nu = 0$ then $n = n/\nu : s = s - 1 :$ goto ...

At each step the positions of ν points are determined. We continue with one position, while the others are saved for later use. This means that for each of the p levels we have to reserve at most $\nu \times p$ memory locations. This does not make heavy demands on working memory.

With the versatile program DUSTB we can make a lot of fractals. The point of departure is always the center $(0,0)$ of **L**, or $(1,0)$ of **R**, which belong to the fractal anyway. This way we can plot up to $2 \uparrow 14 = 16384$ points on an ordinary microcomputer. Compared with the infinite number of points the fractal consists of—a quantity that is not even denumerable—this is hardly any. Still, in practice we get really wonderful images.

With DUSTB we can make fractals for which **L** and **R** are expressed by the following formulas:

$$\mathbf{L} \begin{cases} x' = ax - by \\ y' = bx + ay \end{cases}$$

$$\mathbf{R} \begin{cases} x' = cx - dy + 1 - c \\ y' = dx + cy - d \end{cases}$$

So **L** is a rotation-enlargement around (0,0) with reduction factor $\sqrt{a^2 + b^2}$, and **R** is a rotation-enlargement around (1,0) with a reduction factor $\sqrt{c^2 + d^2}$.

EXAMPLES OF DUSTB. With the choice $a = d = 0$ and $b = c = 0.7$, we get Figure 5.15. Here **L** is a quarter turn with a reduction of 0.7 and **R** is an ordinary central enlargement with respect to (1,0) with a factor 0.7.

Figure 5.15 Fractal generated by a rotation-enlargement and a rescaling

Figure 5.16 Fractal generated by a rotation-enlargement and a rescaling

With the choice $a = b = 0.6$, $c = 0.53$, $d = 0$, Figure 5.16 results. $\mathbf{L}$ and $\mathbf{R}$ are of the same type as in the previous example. Now the rotation angle of $\mathbf{L}$ is $\pi/4$ (45°). The result is reminiscent of a leafy branch. We still see a few discrete points, but if we proceed to an approximation of a higher order, leafy branches will of course start to grow there too on account of self-similarity.

Choosing, finally, $a = 0$, $b = 1/\sqrt{2}$, $c = 1/2$, and $d = -1/2$, we get Figure 5.17. Now both $\mathbf{L}$ and $\mathbf{R}$ are rotations with the same reduction $1/\sqrt{2}$. The angles of rotation are $\pi/2$ (90°) and $-\pi/4$ (−45°) respectively.

Reflection

Line reflection, to be indicated by $\mathbf{S}$, is characterized by an axis of reflection, a symmetry axis (Figure 5.18). Another mirroring restores the original situation. In terms of the algebra of transformations, we can express this by $\mathbf{S}^2 = \mathbf{I}$.

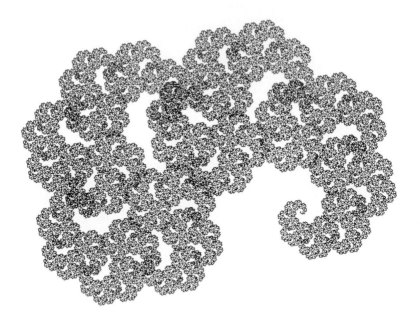

Figure 5.17 Fractal generated by two rotation-enlargements

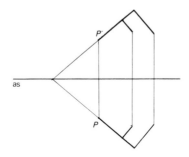

Figure 5.18 Reflection in a line

For fractals, line reflection usually occurs in combination with a change of scale. We often call this combination a contraction-mirroring. Figure 5.19 shows us such a contraction-mirroring. The characteristics of a contraction-mirroring are the center, the only point that stays in

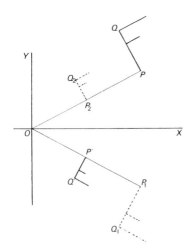

Figure 5.19 Contraction-mirroring

place, and two axes perpendicular to one another, the only lines that stay in place under the transformation. In figure 5.19 these lines are the X-axis and the Y-axis of the Cartesian coordinate system.

In coordinate geometry, contraction-mirroring simply looks like

$$x' = cx$$
$$y' = -cy.$$

This also shows that contraction-mirroring is a reflection in the X-axis ($y \rightarrow -y$) combined with an enlargement with respect to the origin O ($x \rightarrow cx$, $y \rightarrow cy$). The order of the two components does not matter. Thus in Figure 5.19, first mirroring and then multiplying results in the route $P \rightarrow P_1 \rightarrow P'$, whereas multiplying followed by mirroring leads to $P \rightarrow P_2 \rightarrow P'$.

An interesting property of contraction-mirroring is that a repetition is nothing but an ordinary central enlargement. If we apply the formulas above twice, we get $x' = c^2x$, $y' = c^2y$, i.e., the description of a central enlargement with respect to the origin O with a factor c^2.

The most general expression for contraction-mirroring is

(C)
$$x' = ax + by + c$$
$$y' = bx - ay + d$$

with scale factor $\sqrt{a^2 + b^2}$.

These formulas are the counterparts of those of the rotation-enlargement (B). The differences are slight, as they lie only in the sign of y. These formulas are also important in the computer programs.

APPLICATION OF CONTRACTION-MIRRORING. To illustrate contraction-mirroring, we will examine the von Koch curve of Chapter 3 (Figure 3.1). In Figure 5.20 the first couple of phases of that curve's construction are sketched again. Here, too, in every subsequent phase each line-segment is replaced by a meandering line, while the vertices stay where they are. Not surprisingly, these vertices are part of the limit figure that is reached after an infinite number of steps.

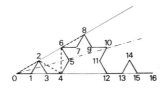

Figure 5.20 Contraction-mirrorings in the von Koch fractal

In Figure 5.20 there are 17 vertices, numbered from 0 to 16. The line that connects the points 0, 2, 4, 6, 8 has exactly the same shape as the line 0, 4, 8, 12, 16 of the first phase, except for the fact that now it is mirrored. Apparently, we are dealing with a contraction-mirroring. From Figure 5.20 we can infer that the axis and the horizontal line 0–16 meet at an angle of $\pi/12$ (15°). The scale factor of the contraction-mirroring is $1/\sqrt{3}$.

Repeating the contraction-mirroring once more, the broken line 0, 2, 4, 6, 8 is transformed into 0, 1, 2, 3, 4. The step from 0, 4, 8, 12, 16 to 0, 1, 2, 3, 4 is an ordinary enlargement with factor 1/3.

In coordinates, contraction-mirroring is described by the system of formulas (C) with $c = d = 0$. If the coordinates of point 16 in Figure 5.20 are $(1,0)$, then the coefficients a and b equal the coordinates of 8: $a = 1/2$ and $b = 1/(2\sqrt{3})$.

So the von Koch curve is invariant under the contraction-mirroring center $(0,0)$. Likewise, there is a second contraction-mirroring center $(1,0)$. We can then express both mirrorings as

$$\mathbf{L} \begin{cases} x' = \dfrac{1}{2}x + by \\ y' = bx - \dfrac{1}{2}y \end{cases}$$

$$\mathbf{R} \begin{cases} x' = \dfrac{1}{2}x - by + \dfrac{1}{2} \\ y' = -bx - \dfrac{1}{2}y + b \end{cases}$$

in which $b = 1/(2\sqrt{3})$.

MORE EXAMPLES OF THE USE OF DUSTB. Using roughly the same computer program, we are now able to design a great variety of fractals, or, more precisely, to approximate them by point clouds. The building scheme is always the same. We start with two transformations **L** and **R**. For the moment, at least, we can select either a rotation-enlargement or a contraction-mirroring.

The fractal is now determined by the smallest number of points that are transformed into one another by **L** and **R** and all combinations that can be derived from them.

If we take the center of **L** as starting point, a point that certainly is part of the fractal, and if we construct all subsequent points by following the principle of the binary tree, then the result will usually be a beautiful figure. Here are some examples.

In Figure 5.21 **L** is a contraction-mirroring with $a = 1/2$, $b = \sqrt{3}/6 = 0.2887$, while **R** is a transformation of the same kind with $a = 2/3$ and $b = 0$.

In Figure 5.22 **L** is a rotation-enlargement with $a = b = 0.4641$, whereas **R** is a contraction-mirroring with $c = 0.6222$ and $d = -0.1965$. The structure of the fractal is highly branched, so this fractal is called a dendrite.

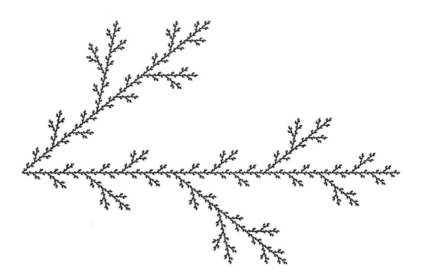

Figure 5.21 Fractal generated by two contraction-mirrorings

Figure 5.23 shows us yet another type of fractal. This one was generated by the two reflections

$$\mathbf{L}\begin{cases} x' = (x+y)/2 \\ y' = (x-y)/2 \end{cases}$$

and

$$\mathbf{R}\begin{cases} x' = (3x-y+2)/5 \\ y' = (-x-3y+1)/5 \end{cases}$$

We can carry out the same construction with three (or more) transformations. In Figure 5.24 we obtained a fractal this way from three rotation-enlargements centered on the vertices of an equilateral triangle. The program for this is called DUSTBT.

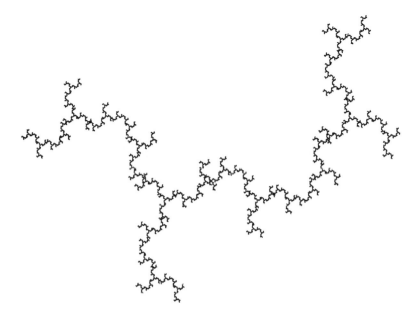

Figure 5.22 Fractal generated by a rotation-enlargement and a contraction-mirroring

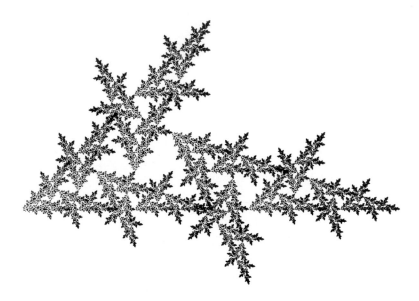

Figure 5.23 Fractal generated by two contraction-mirrorings

Figure 5.24 Fractal generated by three rotation-enlargements

CHAPTER 6

Chance in Fractals

The fractals we have looked at so far have been too beautiful, really, too strictly mathematical. They have had perfect self-similarity, as we can see in the von Koch curve of Figure 3.1. In Chapter 3 we compared that curve with a coastline, like the west coast of Britain. In reality, though, coastlines are created by the whims of nature; chance is at work in this creative process. If we interpret self-similarity not in the exact sense, but statistically, we get more realistic fractals. To do this we require each part of the fractal to have the same statistical properties of form. The mathematical theory underlying such stochastic fractals is far from simple, so we cannot say much about it here.

However, generating these fractals with the computer is quite simple. Computers usually have a built-in chance mechanism, a kind of electronic dice. So we are free to introduce chance into the construction of fractals.

Methods based on chance are often called Monte Carlo methods. A more formal name is stochastic methods. The term stochastics, meaning probability theory and statistics, is derived from a Greek verb for guessing. So stochastics is the science of guessing.

In the present chapter some aspects of chance are discussed under the heading *Monte Carlo*. We will see how fractals can be modified by a kind of controlled randomness.

We will say more about random fractals when we discuss Brownian motion, named after Robert Brown (1773–1858). This biologist observed that microscopic particles follow erratically meandering courses. Nowadays we also speak of *Brownian fractals*. Apart from meandering lines, these include imaginary mountain landscapes. Beautiful Brownian landscapes can be composed using elaborate and costly computer equipment. The resulting landscapes are used for instance in science fiction films.

In a section on Feigenbaum's number we give an account of a recent series of computer experiments that led to spectacular results. These experiments, concerning a simple model of restricted population growth $x \rightarrow ax(1 - x)$, gave new insight into the emergence of chaos from order. The main experiments can even be done on a pocket calculator (preferably a programmable one). Here too we will come across both self-similarity and randomness.

The model of restricted growth dates from 1845, and is usually named after the Belgian mathematician Pierre François Verhulst (1804–1849). It can be extended to an iterative mapping in the plane, with which the names of Julia and Mandelbrot are closely linked. In Chapter 7 we discuss that model at length.

Monte Carlo

There are different ways of bringing chance into the construction of a fractal. But first we must say a few things about the concept of "random number" and about a sequence of random numbers. We also find this term in programming languages, in which the statement "random" (rnd for short) produces a random number.

When we throw a die, any of six numbers 1, 2, 3, 4, 5, or 6 can come up, each equally likely. If we throw many times in succession, we get a completely irregular sequence of random results. However, it is not practical to produce a large sequence of random numbers in this way. In the old days this problem was conveniently solved by using carefully constructed tables. Such tables can be subjected to ingenious statistical tests. The invariable outcome is that no single combination of numbers is predominant. Within statistical limits, all combinations occur equally often. Nowadays, while executing a program the computer can produce an arbitrarily long sequence of random numbers by means of a certain method of calculation. We explain that method in the following example.

We choose an arbitrary four-digit number (the "seed," in computer jargon). This number we square. Next, we remove the first and last digits of the square until we are left with a four-digit number in the middle. We proceed with that number in the same way: square, remove head and tail, and so on. This results in a sequence of numbers between

0 and 9999 that are suitable as a sequence of random numbers. By that
we mean that the sequence does not fail statistical tests of randomness.
On the other hand, the sequence *has* been formed in a regular way.
Everything is fixed by the choice of the first number. So the sequence
is deterministic, but it gives the impression of being chaotic. This is
our first encounter with the concept of "deterministic chaos," a concept
currently receiving much attention in theoretical physics.

In actual fact, at the "rnd" statement the computer forms a group of
zeros and ones that the user can regard as a random number between 0
and 1. Written in decimal, this number has eight digits after the decimal
point. Starting from some initial number, the "seed," the computer
produces a sequence of pseudo-random numbers of arbitrary length
by the method just sketched. This sequence is deterministic; the same
seed will produce an identical sequence. However, as botany teaches
us, not all seeds germinate; similarly in practice this method can fail
from time to time. Fortunately in the great majority of cases it works
all right.

The use of the "rnd" statement can be illustrated very well by a
computer experiment, the result of which reminds us of Mondrian's
paintings. Within a square, horizontal and vertical line-segments are
drawn at random. The position, the length, and the choice of orientation
(vertical or horizontal) of a line-segment depend on chance. Figure 6.1
shows us one drawing made with the program MONDRIAN.

THE ILLUSION OF A COMPLETE FRACTAL. In the previous chapter we
discussed at length the construction of a fractal with a tree structure. If
that structure is binary, the construction is based on the scheme shown
in Figure 5.13. Starting from an initial point P, new points are obtained
by applying two transformations **L** and **R**: first 2, then 4, 8, 16, and so
on. **L** and **R** were similarity transformations, and the resulting fractals
were like the ones pictured in Figures 5.15, 5.16, 5.17, 5.21, 5.22, and
5.23. More generally, **L** and **R** could be other kinds of transformations,
and we will see a number of applications of that in the next chapter.
For now, let us concentrate on the form of the process described above:
each point produced generates two new points. It is like the ancient tale
of the grains of wheat on the chessboard. At each round the number
of points is doubled; it increases exponentially and soon floods the-
memory of the computer. In the previous chapter we demonstrated a

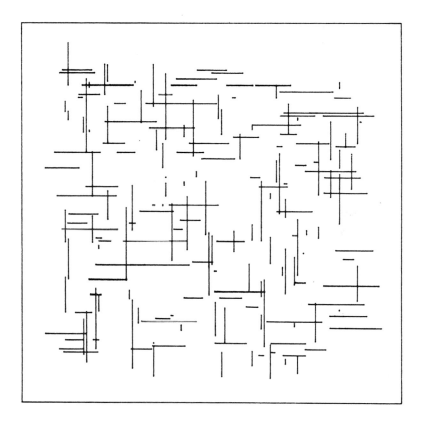

Figure 6.1 Random art à la Mondrian

kind of bookkeeping trick to do more using less memory: the backtrack method.

There is another method, however, based on chance and very effective in practice. The idea is based on the fact that a fractal really consists of an indenumerably infinite number of points and that we can picture only a fraction of them. Ten thousand points may be quite a lot on the screen of a microcomputer, but of course compared with infinity it is nothing at all! The trick is to distribute these points—ten thousand, say—in such a way as to create the illusion of a complete fractal. Up until now we have been doing this very systematically; in the tree of Figure 5.13, for instance, we went twelve levels deep. That way we got $1 + 2 + 4 + \cdots + 2{\uparrow}12 = 8191$ points.

We will now do this differently. We make a random sequence of points: $P_0 \; P_1 \; P_2 \; P_3 \; \ldots$ Each point is derived from the previous one by applying to it either the transformation $\mathbf{L}$ or the transformation $\mathbf{R}$. So $P_1 = \mathbf{L}P_0$ or $P_1 = \mathbf{R}P_0$, or more generally,

$$P_{n+1} = \begin{matrix} \text{either} \; \mathbf{L}P_n \\ \text{or} \quad \mathbf{R}P_n \end{matrix}$$

Whether $\mathbf{L}$ or $\mathbf{R}$ is chosen depends on chance. Using the "rnd" statement we can sum up the above as

$$\text{if rnd} < 1/2 \text{ then } P_{n+1} = \mathbf{L}P_n \text{ else } P_{n+1} = \mathbf{R}P_n$$

APPLICATIONS. We can best illustrate how this works in practice with the aid of a simple example quite familiar to us by now, the Cantor point-set. We will stick closely to what we said about it at the beginning of Chapter 5.

We identify the points of the Cantor set with the numbers between 0 and 1 that can be written in ternary without the digit 1, for instance $x = 0.020222022202$. The transformations $\mathbf{L}$ and $\mathbf{R}$ are

$$\mathbf{L} : x' = \frac{x}{3}$$

$$\mathbf{R} : x' = \frac{2}{3} + \frac{x}{3}$$

In the Monte Carlo method we make a random sequence by tossing dice, as it were:

$$\ldots \mathbf{LRLRRLRRRLR}$$

to be read from right to left! For initial point we choose $x = 0$, the most simple choice. Any other choice would work just as well, though.

The construction can then be read from the following table:

	pattern	x
P_0		0
$P_1 = RP_0$	**R**	.2
$P_2 = LP_1$	**LR**	.02
$P_3 = RP_2$	**RLR**	.202
$P_4 = RP_3$	**RRLR**	.2202
$P_5 = RP_4$	**RRRLR**	.22202
$P_6 = LP_5$	**LRRRLR**	.022202
$P_7 = RP_6$	**RLRRRLR**	.2022202
$P_8 = RP_7$	**RRLRRRLR**	.22022202
$P_9 = LP_8$	**LRRLRRRLR**	.022022202
. . .	. . .	. . .

At each subsequent step the digits shift one place to the right, while on the left either a 0 or a 2 gets added.

This also makes us realize why the choice of the starting point matters so little. For instance, after ten steps the corresponding digits of the expansion of the fraction in ternary have shifted the same number of places. Their contribution is then only on the order of $3\uparrow(-10)$, or 0.000017 in decimal notation.

When we construct a fractal on the screen, numerical accuracy is limited. For the Cantor point-set constructed here, in which line segment [0,1] extends over the entire width of the screen, an accuracy of eight positions (decimal) after the point is quite enough. This means that only $2\uparrow 8 = 256$ points can be pictured. In order to collect these picture points by throwing dice, we will have to toss rather more than 256 times. If we reflect on the statistics of this, we realize that for every 1000 random choices, five points on average still elude us. We have to go up to at least 1600 random choices to get just about all of them.

Everything said about the Cantor point-set applies in principle to any fractal constructed with the Monte Carlo method. How long we have to go on throwing dice is usually a matter of experimenting within the limits of computing speed and screen resolution.

We made the fractal of Figure 6.2 using a version of the program DUST. The transformations are

$$\mathbf{L} \begin{cases} x' = -y \\ y' = x \end{cases}$$

$$\mathbf{R} \begin{cases} x' = 1 + a(x-1)/\left((x-1)^2 + y^2 + 1\right) \\ y' = ay/\left((x-1)^2 + y^2 + 1\right) \end{cases}$$

with $a = 2.8$.

Figure 6.2 Fractal generated by two transformations on the basis of the Monte Carlo method

L is a quarter turn around (0,0). **R** is an enlargement with respect to (1,0) with a variable factor that depends on the distance of (x, y) from the center (1,0). The figure contains a number of symmetries, of which the program takes full advantage. The calculation of a single point simultaneously yields the positions of seven more.

A MODEL FOR NATURAL OBJECTS. Introducing chance as small disturbances while constructing fractals, we can make fractals serve as a model for natural objects like trees, plants, sponges, and corals. To illustrate this, we will construct a kind of windblown Pythagoras tree (Figure 6.5). First we look at Figure 6.4, a variation of the Pythagoras tree of Figure 4.14. The basic motif is the trunk, with two places AB and BC for attaching the side branches (Figure 6.3). In the program for this, PYTHTD, the positions $(0,3)$, $(0.5,3.5)$, and $(1,3)$ were chosen for A, B, and C.

Figure 6.3 Trunk of a Pythagoras tree

As we know, the construction comes down to the trunk motif being copied in similar form on a diminishing scale. Before attaching the scaled-down motif, however, we will change it slightly in a random manner. In the program this can be done for instance by changing the parameter a as follows:

$$a \rightarrow a + w(\text{rnd} - \tfrac{1}{2})$$

The number in brackets is a random number between -0.5 and $+0.5$. With the factor w we influence the intensity of the random disturbances. If $w = 0$, chance is eliminated and Figure 6.4 results. If on the other hand $w = 0.1$, we get the windblown tree of Figure 6.5.

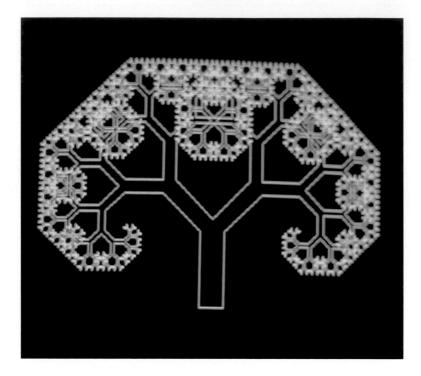

Figure 6.4 Mathematical Pythagoras tree

Brownian Motion

In 1828 the Scottish biologist Robert Brown (1773–1858) discovered a peculiar phenomenon. While looking through his microscope at small particles floating in a liquid, he was struck by the fact that the particles made tiny, erratic, unpredictable movements. He put this phenomenon down to physical causes. The theory behind this Brownian movement was understood only much later, in 1905—by Einstein, no less. The molecules of the liquid make irregular thermal motions that get more vigorous as the temperature rises. The molecules continually bump against the much bigger particles that can be seen through the microscope. This causes these particles to move erratically.

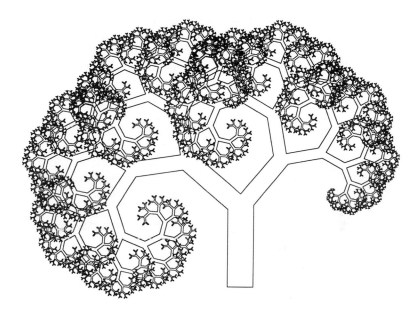

Figure 6.5 Randomly deformed Pythagoras tree

The concept of "fractal," or rather fractal curve, helps us to form an impression of the trajectory of a particle undergoing Brownian motion. That trajectory is like a three-dimensional coastline with meanders on every scale. The principle of self-similarity still holds, provided we attach a probabilistic clause to it. It comes down to the fact that the *statistical* properties of each part, however small, do not depend on the scale. We can then paraphrase the result of Robert Brown's experiment by stating that through his microscope Brown was looking at a spatial fractal curve. He could not see the smallest meanders at the molecular level; the larger ones, however, which came within the reach of his microscope, he did see. With a microscope that was ten or a hundred times more powerful, he would have seen virtually the same thing!

Following Mandelbrot, we can now extend the concept of "fractal" to a geometrical structure with statistical self-similarity. Consequently the path of the microscopic particle is a Brownian fractal curve. Analogously, there also exists a Brownian surface or Brownian landscape.

In some ways, Brownian fractals are much more natural. A lunar landscape could very well be seen as a fractal surface. A few large craters
represent the largest scale. On nearly every smaller scale, craters can
also be seen. The size and location of the craters depend purely on
chance; in short, we have a fractal moon. Things it took nature hundreds of millions of years to create can now be imitated by a powerful
computer within minutes.

Even though the mathematical principles, a sequence of geometrical
transformations based on chance, are quite simple, this demands rather
a lot of the computer technically. With the aid of the computer a realistic image, a projection, has to be made of the object that has been
determined geometrically—a lunar landscape or a planet in a science
fiction film, say. Preferably this should be done in color, with direc-
tional lighting, and only the parts visible to an imaginary observer
should be pictured. This technique, computer graphics, is of growing importance in the world of film, television, and industry. It does

Figure 6.6 Brownian landscape. By Richard E. Voss, from *The Fractal
Geometry of Nature* by B. B. Mandelbrot (W. H. Freeman),
copyright © 1982 by Benoit B. Mandelbrot

require elaborate and costly computer apparatus. Figures 6.6 and 6.7, taken from Mandelbrot's book *The Fractal Geometry of Nature*, give us an idea of the possibilities.

Figure 6.7 Brownian landscape. By Richard E. Voss, from *The Fractal Geometry of Nature* by B. B. Mandelbrot (W. H. Freeman), copyright © 1982 by Benoit B. Mandelbrot

BROWNIAN LINE. With a microcomputer costing a hundred times less, we cannot do so much, of course, but it is still an excellent device for giving an idea of the mathematical principles involved. Our aim will be modest: to construct the fractal of a Brownian line in the plane. Here we can think of a coastline or the elevation of the land along a straight line, from Cologne to Paris, say. The easiest way to get such a Brownian line is as follows.

We take a line-segment and divide it into, say, 1001 points: $x_0 = 0$, $x_1 = 0.001$, $x_2 = 0.002$, ..., $x_{1000} = 1$. With the computer we produce a sequence of a thousand random numbers $r_1, r_2, r_3, \ldots, r_{1000}$ between 0 and 1. Next we draw a graph. With a suitable scale factor we plot, for the kth point, the value of $(r_1 - 1/2) + (r_2 - 1/2) + (r_3 - 1/2) + \cdots + (r_k - 1/2)$ against $x_k = k/1000$. In this way we get a sequence of points $P_0, P_1, P_2, \ldots$ in which P_0 is the left-most point of the original line-segment. Since for each k, $(r_k - 1/2)$ is between $-1/2$ and $+1/2$, each point lies by chance either a bit higher or a bit lower than the one before. The result, obtained with the program BROWNL, is Figure 6.8.

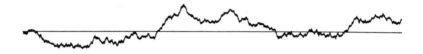

Figure 6.8 Brownian curve of winnings and losses

This line can be interpreted as the graph of the winnings and losses of two players gambling with each other. If Peter wins, then Paul will pay him a fixed amount. If Peter loses, he pays Paul the same amount. The game is repeated many times—a thousand, say. After each game Peter writes down his total winnings or Paul's total losses, which comes down to the same thing. This game is worth playing on the computer. The result is a different Brownian curve each time. Sometimes Paul keeps winning; on other occasions he keeps losing. However, fortune takes turns in smiling on them.

Feigenbaum's Number

There is no Nobel prize for mathematics. This is a pity, because it means that spectacular mathematical discoveries do not get the publicity they deserve. In this section we will discuss a recent discovery, once again concerning self-similarity repeated to infinitely small scales. It is based on a numerical experiment any student can do with a simple programmable pocket calculator. The subject is a mathematical model of restricted population growth, which, for instance, describes numbers of insects (flies, say) in successive generations. We apply some scaling here, so that the number of flies x is always a number between 0 and 1.

The model for *unrestricted* growth is very simple: $x \rightarrow ax$. This means that in each generation there will be a times as many flies as in the generation before. If for instance $a = 2$ and we start with $x = 0.001$, then we get the sequence

$$x_0 = 0.001, \quad x_1 = 0.002, \quad x_2 = 0.004, \quad x_3 = 0.008,$$
$$x_4 = 0.016, \quad x_5 = 0.032, \quad x_6 = 0.064, \quad x_7 = 0.128,$$
$$x_8 = 0.256, \quad x_9 = 0.512.$$

The suffix counts the generations. Since we supposed that $0 \leq x \leq 1$, we have to stop at the ninth generation. After all, if we went on we would get $x_{10} = 1.024$, $x_{11} = 2.048$, and so on. Now the model is no longer realistic!

The well-known economist Malthus (1766–1834), who devoted much time to studying models of growth, is remembered in the name of the factor a: the Malthusian factor. This number can be interpreted as the degree of fertility of the insect population. In 1845 P. F. Verhulst derived a model of restricted growth from this by supposing that the Malthusian factor decreases as the number x increases. The biggest population the environment will support is $x = 1$. If there are x insects, $1 - x$ is a measure of the space nature permits for population growth. Consequently we replace a by $a(1 - x)$. The model then becomes $x \rightarrow ax(1 - x)$, or, using indices,

(A) $\qquad x_{n+1} = ax_n(1 - x_n)$.

If we use the same example with $a = 2$ and $x_0 = 0.001$, we get

$$x_0 = 0.001, \quad x_1 = 0.002, \quad x_2 = 0.004, \quad x_3 = 0.008,$$
$$x_4 = 0.016, \quad x_5 = 0.031, \quad x_6 = 0.060, \quad x_7 = 0.113,$$
$$x_8 = 0.201, \quad x_9 = 0.321, \quad x_{10} = 0.436, \quad x_{11} = 0.492,$$
$$x_{12} = 0.500, \quad x_{13} = 0.500, \quad \dots$$

We notice that in the beginning there is hardly any difference, but from the seventh generation on, the check on growth becomes more and more apparent. Eventually the number becomes constant. At that point a biological balance between population and environment has been reached.

Realistic biological models are usually more complicated, of course. Still, this model (A) has nearly all the characteristics of more intricate models. The model is often called Verhulst's model. It has been thoroughly studied and discussed in the scientific literature. Though this discussion mainly takes place on the peaks of the highest mathematical mountains, where beautiful views are the prerogative of specialists, we can all get a lot of pleasure out of it, though we will have to do without their deeper insights. So, for the moment, let us descend into the lowlands, where we will first make a few simple observations about Verhulst's model.

To start with, we detach the model from its biological background, and look at it through the eyes of a mathematician. If x is a number between 0 and 1, then $x(1 - x)$ will be maximally 1/4 and $ax(1 - x)$ maximally $a/4$. So if we limit the factor a to the interval [0,4], then we can be certain that the transformation $x \rightarrow ax(1 - x)$ will not take us out of the interval [0,1]. So we suppose $0 \le a \le 4$ and do the following experiment. After choosing a fixed value for a and an arbitrary initial value x_0, we produce a sequence $x_1, x_2, x_3, \ldots$ on the computer. We then ask ourselves what happens in the long run. We will do this systematically. If $0 < a \le 1$, x_n approaches zero, so $x_n \rightarrow 0$. In the biological model this means the insects die out if they are not fertile enough.

If $1 < a \le 2$, then x_n turns out to approach an equilibrium value. This limit can be calculated quite easily by replacing both x_n and x_{n+1} in (A) by the limit value L. From the equation $L = aL(1-L)$, the result $L = 1 - 1/a$ follows immediately. What is more, the sequence $x_1, x_2, x_3, \ldots$ turns out to contain either increasingly large or increasingly small numbers. Mathematicians say this sequence is either increasing or decreasing.

If $2 < a \le 3$, experiments show us that the sequence approaches a limit value $1 - 1/a$, only now the limit is approached from both sides. This is called oscillation. In the biological model it means that if $1 < a \le 3$ there is a stable equilibrium. There is a balance in nature.

ONE MORE STEP. We will now carefully advance one more step and take $a = 3.2$. After a while the values end up oscillating between 0.5130 and 0.7995. Yet just now we found that the equilibrium value would be $1 - 1/a = 0.6875$. This is indeed correct, but now the equilibrium is unstable, which means it is susceptible to small disturbances. It is rather interesting to start the sequence with $x_0 = 0.6876$, very close to equilibrium. It then turns out that we slowly drift away from the equilibrium value and end up in the 2-cycle of 0.5130 and 0.7995.

If we increase a a bit more, this phenomenon is repeated. If, however, a passes the value 3.4495 ($1 + \sqrt{6}$, to be precise), the 2-cycle also becomes unstable and a 4-cycle appears. If $a = 3.5$, for instance, and we start with $x_0 = 0.2$, then after 25 steps we are in the 4-cycle to an accuracy of four decimals: 0.5009, 0.8750, 0.3828, 0.8269.

At $a = 3.5441$ something happens again. The 4-cycle stops being stable and turns into an 8-cycle. The values of a for which these transitions from one cycle to another occur, are called bifurcation points (the Latin *furca* means fork); the transitions are called *bifurcations*. The phenomenon is called period doubling.

GREAT DISCOVERIES. We are about to make some great discoveries. First let us summarize our findings to date in the following table:

m	a	increase in a	ratio of consecutive increases
1	3	—	—
2	3.449499	0.449499	—
3	3.544090	0.094591	4.75
4	3.564407	0.020313	4.66
5	3.568759	0.004352	4.67
6	3.569692	0.000933	4.67
7	3.569891	0.000199	4.7
8	3.569934	0.000043	4.6

Alongside the index m we indicate the bifurcation value of a at which a cycle of $2 \uparrow m$ elements comes into being. The a-values increase in ever smaller steps. These we find in the third column. The physicist Mitchell Feigenbaum was struck by the fact that this sequence a_1, a_2, a_3, ... approaches a certain limit value more or less like a geometric sequence. This he inferred from the fact that the difference between two successive a-values decreases by about the same factor, 4.669, every time. A very accurate calculation, on the computer, of course, gave the value

$$F = 4.6692016\ldots$$

In itself this was not all that remarkable, so initially Feigenbaum paid little attention to it. Some time later, however, while researching an entirely different model, $x_{n+1} = a\sin(\pi x_n)$, he found not only the same phenomenon, but also exactly the same F-value. In brief, it became obvious that this was a *universal* phenomenon that always occurs whenever there is repeated period doubling. F is a universal constant, just like π or e, the base of the natural logarithms. F has been called *Feigenbaum's number* ever since.

Once this became generally known, the same number was found in various experiments in physics that all have a so-called phase transition in common. The behavior of helium near absolute zero is an example of this. All this demonstrates beautifully that seemingly pure mathematical research, computer experiments, and physical reality are in fact intimately related.

ORDER AND CHAOS. The above phenomenon of period doubling in the model $x \rightarrow ax(1 - x)$ reminds us of a binary-based fractal with a scale factor roughly equal to Feigenbaum's number. The period doubling in the model (A) stops at the value $a_\infty = 3.569946$. One wonders what will happen after that, until $a = 4$. Well, it is enough to fill a book. So we will have to restrict ourselves severely here. In short, for $a_\infty < a \leq 4$ the model will behave either chaotically or periodically. Thus there is a tiny interval near $a = 3.83$ where a stable cycle of three elements occurs, a so-called window of periodicity.

If we let a increase a little up to the value 3.841, then that cycle gets unstable and changes into a stable 6-cycle: another period doubling. This is the start again of a whole range of doublings 3, 6, 12, 24, ... just like the ones we saw before. Everything repeats itself on a smaller scale. In actual fact, this period doubling is also governed by Feigenbaum's number. Between a_∞ and 4 there are many more windows for which the model is periodic, a denumerably infinite number of them, even. Still, an indenumerable number of values of a remain for which the model behaves chaotically.

Figure 6.9 shows us this varied behavior.

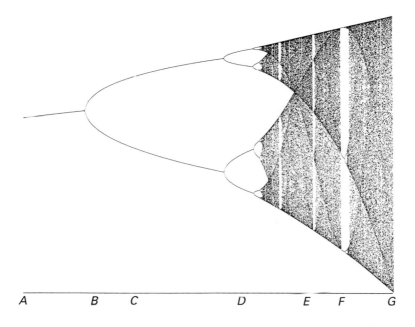

Figure 6.9 Order and chaos in the model of restricted growth $x \rightarrow ax(1-x)$

A	$a = 2.9$	single limit point
B	$a = 3.0$	transition to 2-cycle
C	$a = 3.2$	2-cycle
D	$a = 3.5$	4-cycle
E	$a = 3.74$	5-cycle
F	$a = 3.83$	3-cycle
G	$a = 4.0$	complete chaos

Horizontally, the parameter a has been plotted from 2.9 to 4. Values x_n of the iterative process (A) have been plotted vertically. Here n is sufficiently large for the values to be nearly those of the situation in the limit when $n \rightarrow \infty$. For $2.9 \leq a \leq 3$ this gives the simple equilibrium $1 - 1/a$; for $3 \leq a \leq 3.4495$ a 2-cycle arises; after that a 4-cycle, and so on. Since we let a increase in small steps, some detail is lost. Further on, however, the large window with a stable 3-cycle can be seen clearly, and we also see a smaller window with a stable 5-cycle.

The figure has a high degree of self-similarity. Let us look, for instance, at the region near one of the points of the stable 3-cycle. With roughly the same computer program (COLLET, named after one of the first researchers), Figure 6.10 shows the stretch $3.835 < a < 3.855$. Everything we saw in Figure 6.9 recurs on a more reduced scale. We can carry on this way indefinitely. We would have observed the same at any other element of a stable cycle!

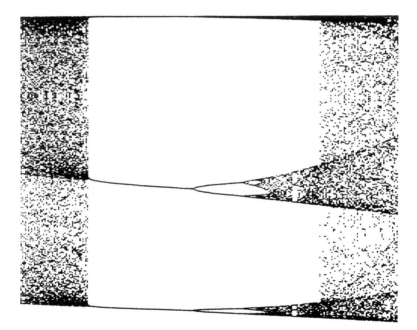

Figure 6.10 Detail of the previous diagram around $a = 3.83$.
Here a 3-cycle arises and starts bifurcating

Illustrations such as Figures 6.9 and 6.10 are especially useful as a graphical computer experiment to quickly get both an understanding and a survey of the peculiarities of an iterative model of the type $x \rightarrow f(x)$, of which (A) is a special case. These figures can be considered forerunners of the Mandelbrot fractals, which will be discussed in the next chapter.

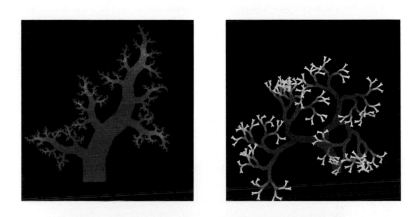

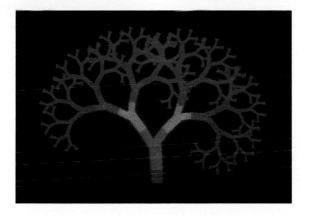

Fractal models of sponges and corals

CHAPTER 7

Poincaré, Julia, Mandelbrot

This chapter takes us among the mountain peaks in the land of fractals. One panoramic view after another will unfold. Our guide will be Mandelbrot, who has exerted himself with never-ending enthusiasm to bring fractals to the notice of mathematicians, physicists, and artists. His remarkable discoveries and his computer experiments have inspired many people; today a particular kind of fractal bears his name. This chapter will lead to a series of fantastic color shows, provided one has a powerful computer. The self-similarity of a mathematically defined fractal makes it possible to magnify arbitrarily small details into color pictures in their own right. Mandelbrot and mathematicians such as H.-O. Peitgen and Sh. Ushiki have made spectacular use of this possibility. Here we reach the frontier between mathematics and art. The starting point is a mathematical model; choosing the detail and selecting the colors belong to the domain of art.

Though born in Poland (1924), Mandelbrot was educated in France, so it is no coincidence that many of his computer experiments elaborate on the work of his compatriot Gaston Julia (1893–1978). In 1919 Julia published a prizewinning study in which he founded the mathematical theory of a type of fractal generated by a so-called iterative conformal transformation. A conformal transformation is a transformation that leaves angles unchanged. The principal example of this is

$$(A) \quad \begin{aligned} x' &= x^2 - y^2 + a \\ y' &= 2xy + b. \end{aligned}$$

Here a and b are arbitrary numbers. For every value of a and b we get a fractal! Fractals like that are nowadays called Julia fractals. It is quite simple to make them appear on the screen of a microcomputer, if we do not demand too much resolution and make only modest use of color. In the high mountains one can still take beautiful photographs in black and white with a simple camera.

Julia fractal, type (A)

Many lines of research emanate from another Frenchman, Henri Poincaré (1854–1912), one of the last universal mathematicians. With his work on the motion of (among other things) the celestial bodies, he founded the theory of what is now called dynamical systems. Concepts like deterministic chaos and self-similarity typify this wide research area. As far as fractals are concerned, iterative mappings in the plane— so-called Poincaré mappings—are of special importance. The fact that this is a dynamical system means that these mappings are conservative or measure-conserving, i.e., a small circle is transformed into a small ellipse of the same area.

In 1969 the French mathematician and astronomer M. Hénon

studied what may well be the most simple mathematical model of its kind:

(B)
$$x' = y$$
$$y' = -x + 2ay + by^2$$

This iterative mapping serves as a model for a wide range of physical phenomena, from the motion of celestial bodies to the interaction of elementary particles. A great number of computer experiments showed Hénon and his followers different aspects of self-similarity and chaos—in short, fractal-like structures. Poincaré had predicted their existence, but only with the aid of modern computer technology could they be made visible. Here too we are nearing the frontier between mathematics and art. With models of the type (B), fractal-like point clouds, fractal dust, can be created. French researchers call this "aesthetic chaos." We will give a number of examples of this.

With a microcomputer, an approximation to most if not all of the illustrations in this chapter can be made without trouble. With the programs of Appendix B, many more fractals can be made by choosing different parameter values. Our final chapter is intended for the do-it-yourself enthusiast. It gives some variants of (B) with which one can experiment to one's heart's content.

Celestial Mechanics

Ever since the beginning of human civilization, man has let his activities on earth be guided by the orbits of the celestial bodies: the sun, moon, planets, stars, and comets. At the beginning of the Christian era, astronomy had evolved into an impressive science. A large number of observations could now be described and summarized by mathematical methods and numerical tables. People knew the earth was round, and they gave fair estimates of both its radius and the distance between it and the moon.

In the second century Greek science came to an end with the work of Claudius Ptolemaeus (90–168), in some respects a precursor of Newton! Ptolemy invented an accurate though rather complicated theory for explaining the orbits of the planets as observed from earth. One would have to imagine clockwork on a cosmic scale, in which a large number of cogwheels, big ones, small ones, and tiny ones, interact with one another. Ptolemy's theory is known as the epicyclic theory: cycles upon cycles upon cycles ... In this we sense the principle of self-similarity, a cosmic motion with the structure of a fractal!

Thanks to the emergence of the Arabic culture, the scientific legacy of the Greeks was preserved. Ptolemy's work lived on as the *Almagest*, the textbook on astronomy throughout the Middle Ages. In 1543 a new era began with Nicolaus Copernicus (1473–1543). By putting the sun in the center he was able to simplify Ptolemy's theory, reducing it from 77 epicycles to 34.

The next step was made by Johannes Kepler (1571–1630) about fifty years later. After many attempts based on the accurate observations of the Danish astronomer Tycho Brahe (1546–1601), he discovered that the planets move around the sun in elliptical orbits with the sun as focus.

It was left to Isaac Newton (1642–1727) to give the work of his famous predecessors a grand finale. When he was about twenty-four he formulated the law of gravitation. He proved that all objects, from small to large, from a grain of sand to a planet, are subjected to one and the same universal law. At the same time he developed the mathematical methods needed to describe the orbits of celestial bodies with mathematical equations. Newton was then able to explain theoretically the laws that Kepler found solely by experimentation.

At the end of the last century Henri Poincaré made a thorough study of "mécanique céleste," or celestial mechanics. He pointed out that even though Newton's theory is deterministic, the motions of celestial bodies attracting one another can be very complicated, complicated to such an extent that in the long run their behavior is unpredictable and even chaotic. The situation is completely analogous to the deterministic chaos we mentioned in the previous chapter when we discussed the model $x \rightarrow ax(1-x)$. With this study Poincaré laid the foundation for *dynamical systems*, as they are called today.

POINCARÉ'S MODEL. Poincaré and many others after him have made it clear that much insight into the sometimes rather complicated behavior of dynamical systems can be obtained from quite simple mathematical models. Poincaré already suspected the existence of a fractal-like

structure. More than half a century had to pass, however, before technology had progressed so far that these fractals could really be made
visible on the screen of a computer.

Here we will look at one type of model only, an iterative transformation of points in the plane:

(C)
$$x_{n+1} = f(x_n, y_n)$$
$$y_{n+1} = g(x_n, y_n).$$

In this f and g are prescribed functions.

We came across such iterative transformations while dealing with
rotations, mirrorings, etc. Thus:

(D)
$$x_{n+1} = x_n \cos \alpha - y_n \sin \alpha$$
$$y_{n+1} = x_n \sin \alpha + y_n \cos \alpha$$

is the model of a rotation in which at each iterative step the image
point (x_n, y_n) rotates around the origin through the angle α. If P_0 with
coordinates (x_0, y_0) is the starting point, then the subsequent points P_1,
P_2, P_3, ... will be situated together with P_0 on a circle around the
origin. The point-series P_0, P_1, P_2, ... is called the orbit of P_0.

Under certain circumstances the orbit will close on itself and reduce
to a limited number of points, a so-called periodic cycle. If, for instance, $\alpha = \pi/3$, then a cycle of six points will come into being. If
$\alpha = 5\pi/7$, then we go around the circle exactly five times in fourteen
steps. Whenever α/π is a common fraction (a rational number), the
result will be a periodic cycle. If, however, α/π is irrational, then the
orbit of an arbitrary point will consist of an infinite number of points
lying on a circle completely filled by those points.

In the models (C) looked at by Poincaré, the transformation $x \rightarrow$
$f(x, y)$, $y \rightarrow g(x, y)$ is area-conserving. This means that an arbitrary
circle is transformed into a closed curve of the same area.

HÉNON'S MODEL. In 1969 M. Hénon studied the iterative model

(E)
$$x_{n+1} = ax_n - b(y_n - x_n^2)$$
$$y_{n+1} = bx_n + a(y_n - x_n^2).$$

Here $a = \cos \alpha$ and $b = \sin \alpha$. In the introduction to the present

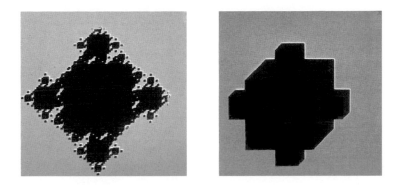

A cubist reconstruction of a spatial fractal

chapter this model was presented as the plane mapping (B), which at first sight looks very different. A small calculation, however, shows that (B) and (E) are equivalent. They differ only in the definition of the y-coordinate, and by a scale transformation. Anyone who wants to check this must replace the expression $ax_n - b(y_n - x_n^2)$ in (E) by ν_n, which comes down to defining a different y-coordinate, given by ν. Mathematicians will not have any difficulty deriving version (B) from (E) in this way, nor will they find it difficult to verify that the model (E) is area-conserving.

Without calculation we notice that (E) differs only slightly from the rotation (D) if x_n is so small that the square x_n^2 is negligible compared with the linear terms in x_n and y_n.

In Figure 7.1 we show one of Hénon's results made with the program HENON. We have drawn a number of orbits for the case $a = 0.24$, $b = 0.9708$, which corresponds to $\alpha = 76.1135°$. This means going around once in just under five steps.

In Figure 7.1 we can discern different kinds of orbits. The origin O is an equilibrium point and so forms an orbit that consists of one single point. Around it we see circular orbits in which the point-series almost links up into a closed curve. The curve is either everywhere densely covered with points of the orbit or reduces to a periodic cycle.

The five nests of closed curves, an island structure, strike us immediately. The number five is of course linked to the fact that near the origin the iteration differs only slightly from a fifth of a turn. The

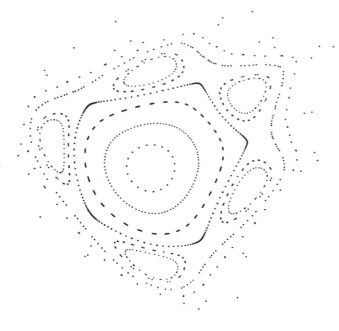

Figure 7.1 Hénon's experiment

centers of these five groups of islands form a periodic cycle of five elements themselves.

Finally we notice an irregular external orbit, a so-called chaotic orbit.

INTERPRETATION. The origin can be interpreted as a point of stable equilibrium. This means that an orbit that starts near this equilibrium point will stay near it, and in this case will lie on an almost circular curve. The same goes for the centers of the periodic 5-cycle. At each iteration step of (E) they shift one position, and after five steps they are back where they started. If we indicate the "near-rotation" symbolically by E and interpret E^5 as the fivefold repetition of this, then the five periodic points we mentioned have become ordinary equilibrium points, just as the origin was an ordinary equilibrium point for E. This is the first indication of the self-similarity inherent in the figure. This self-similarity is characteristic of this type of dynamical system.

Let us now look at the island structure around each of these five points. If we move away from those points very carefully, we see that at each subsequent starting point of an orbit the islands increase in size until they touch one another. This situation is sketched in Figure 7.2. Here I_0, I_1, I_2, I_3, and I_4 are the ultimate points of tangency of the largest islands. They too form a periodic cycle. The centers of the islands—the smallest islands, in fact—form the periodic cycle S_0, S_1, S_2, S_3, S_4. This S-cycle is stable. An orbit starting in S_0, for example, will pass near S_1, S_2, S_3, and S_4 and will return close to S_0. It goes on like this, and the result is either a set of five islands that again can be filled densely with points, or a cycle of a higher periodicity.

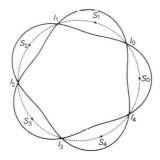

Figure 7.2 Scheme of a structure with five islands

The I-cycle, on the other hand, is unstable. Both theory and experiments show that an orbit that starts in one of the unstable points, I_0 for instance, is chaotic. This chaos can be compared to the deterministic chaos we discussed in the previous chapter when looking at the model $x \rightarrow ax(1 - x)$. Figure 7.3 shows a part of the chaotic orbit starting at the point $x_0 = 0.57$, $y_0 = 0.16$, very close to I_0. The picture speaks for itself. The fact that it is virtually impossible to predict what will happen next is characteristic of a chaotic orbit like this. Rounded to four decimals, the coordinates of I_0 are $x_0 = 0.5696$, $y_0 = 0.1622$. After a hundred iteration steps we find $x = 0.5386$, $y = 0.1864$. This is still quite close. Had we, however, started with $x_0 = 0.57$, $y_0 = 0.16$,

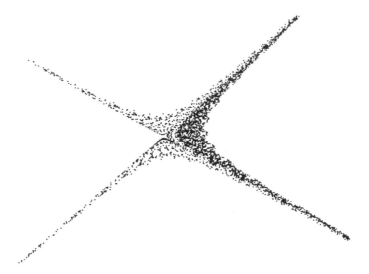

Figure 7.3 Detail of a chaotic orbit in Hénon's experiment

we would have gotten $x = 0.3176$, $y = 0.4202$, somewhere quite different. This shows us the main characteristic of deterministic chaos: things that in the beginning are very close together will drift further and further apart in the long run.

Poincaré's theory says that Figure 7.1 is full of ring-shaped structures like those of Figure 7.2. Stable periods and unstable ones alternate. The stable equilibria are surrounded by stable orbits—islands. The unstable equilibria are surrounded by chaotic orbits. We find that in many cases the chaotic orbits have a ring-shaped structure, and are themselves bordered by stable orbits. In Figure 7.3 it can indeed be seen that the chaotic orbit is surrounded by parts of large islands that are not pictured themselves.

So the structure generated by **E** consists of periodic cycles that are either stable or unstable: stable orbits that fill up a closed curve and unstable chaotic orbits. That structure is universal and in principle applies to every area-conserving transformation (C).

We already have the beginnings of self-similarity with the transition from **E** to $\mathbf{E}^5$, the fivefold repetition of the mapping **E**. For $\mathbf{E}^5$

each large island in Figure 7.1 behaves like an independent structure of roughly the same character as the entire figure. S_0 for instance now plays the part of the origin. All we said about the entire figure also applies to each of the five islands. So once again we have order, chaos, and smaller island structures. Clearly we can continue like this endlessly, with an ever-repeating microstructure of islands containing islands containing ... We can make parts of that microstructure visible using the computer, of course, by picturing one or more chaotic orbits. This shows us a way of making beautiful pictures on the computer. One "thinks up" a model of the type (C) in which f and g must be such that the transformation is area-conserving, and one experiments with the microcomputer.

A useful way of getting a good model is based on a so-called two-step iterative model:

(F) $x_{n+1} + x_{n-1} = 2F(x_n)$.

Here $F(x)$ is an arbitrarily chosen function of the variable x. We can explain the significance of such a model by looking at Fibonacci's model. This describes the unchecked growth of, say, a plague of rabbits.

The number of female rabbits per generation follows the series

$$1 \quad 1 \quad 2 \quad 3 \quad 5 \quad 8 \quad 13 \quad 21 \quad \ldots$$

If we take $x_0 = 1$, $x_1 = 1$, $x_2 = 2$, $x_3 = 3$, $x_4 = 5$, and so on, then $x_{n+1} = x_n + x_{n-1}$. This gives us the general rule that the number at "time" $n + 1$ is determined by the numbers at both "time" n and the earlier "time" $n - 1$. By the way: in this special case an exact expression for x_n can be derived together with an approximate formula $x_n \approx (1.618)^{n+1}/\sqrt{5}$, where $1.1618 \ldots$ is the golden section $(-1 + \sqrt{5})/2$.

From (F) we can derive an iterative mapping in two dimensions by making $x_{n+1} = y_n$ and interpreting the second variable y_n together with x_n as the coordinates of a point in the plane.

The new form of (F) is then

(G)
$$x_{n+1} = y_n$$
$$y_{n+1} = -x_n + 2F(y_n)$$

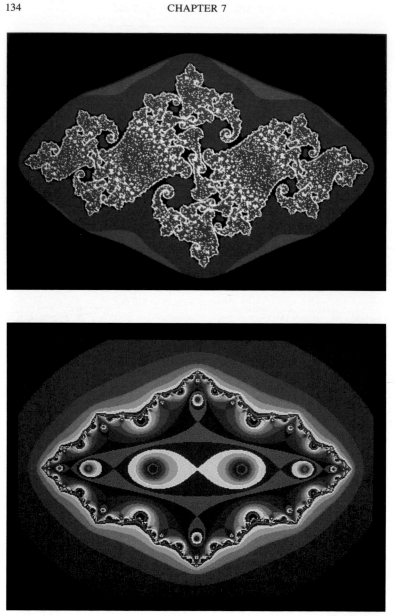

Julia fractals, type (A)

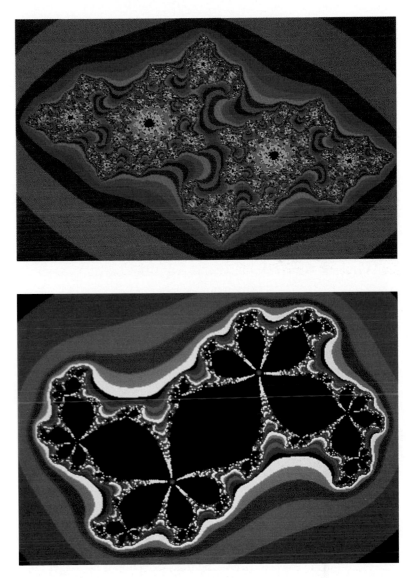

Julia fractals, type (A)

GUMOWSKI AND MIRA. Physicists go to great lengths to discover the secrets of nature by making elementary particles interact with one another. Experiments of this kind are carried out in Geneva at CERN, a European research institute. They make heavy demands on the equipment. One has to make elementary particles like protons move at high speeds in an accelerator, a circular channel with the diameter of a tin can but several kilometers long. The orbits of the particles have to be stable. We can see that a great many problems arise here, both technical and mathematical. The physicists (or mathematicians) Gumowski and Mira examined a model of the type (F) or (G). In their books and articles they give a great many illustrations of computer experiments, including many examples of "chaos esthétique." For their main model they choose

$$F(x) = ax + \frac{2(1-a)x^2}{1+x^2},$$

in which a is a parameter to be chosen freely.

For $a = 1$, $F(x)$ reduces to x. The model (G) is then completely regular, all is stable, and chaos is absent. As a deviates more from 1, more chaos will occur and the computer pictures get more interesting. Below we give a couple of examples based on the program MIRA. Instead of (G) we use a slightly different but equivalent scheme:

(H)
$$\begin{aligned} x_{n+1} &= y_n - F(x_n) \\ y_{n+1} &= -x_n + F(x_{n+1}) \end{aligned}$$

This formulation generates a fractal structure symmetrical with respect to the X-axis.

Figure 7.4 shows us a single chaotic orbit of the model (H) for $a = 0.31$ with starting point $(12, 0)$, or rather the first few thousand points of it.

If in the second equation of (H) we replace the term $-x_n$ by $-bx_n$ with $b = 0.9998$, the result will be Figure 7.5. That minute difference has dramatic consequences. The transformation is no longer area-conserving; areas are reduced by the factor b. The effect of this is that the orbits tend to spiral inward. In Figure 7.5 we therefore pictured one single orbit starting from the point $(15, 0)$ for the case $a = 0.7$.

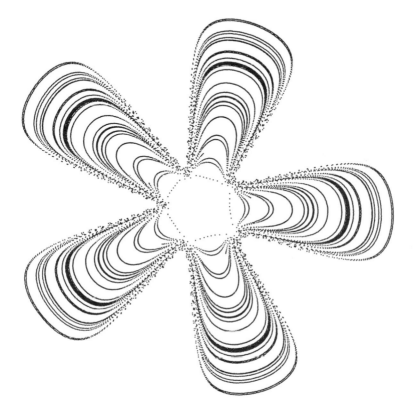

Figure 7.4 Aesthetic chaos in Mira's model

STRANGE ATTRACTOR. Iterative processes of the type (C) that are not area-conserving can sometimes lead to beautiful fractal-like pictures. These pictures do not depend on the choice of starting point, and nearly all orbits seem to lead to the same limit figure. Mathematicians call this a "strange attractor." The word "strange" reflects how amazed they were, even though as mathematicians they are accustomed to a lot of strange things, when they first saw pictures like this appear on

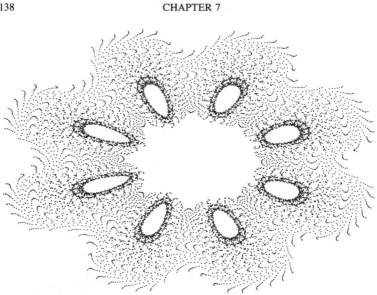

Figure 7.5 An orbit spiraling inward in Mira's model

the screen. Figure 7.6 shows just one example of this. The iterative mapping is

$$x_{n+1} = (2 - a)x_n - by_n$$

$$y_{n+1} = -bx_n + ay_n$$

for $|x_n| < 1/2$ and

$$x_{n+1} = ax_n - by_n + (1 - a)\operatorname{sgn} x_n$$
$$y_{n+1} = bx_n + ay_n - b \operatorname{sgn} x_n$$

for $|x_n| \geq 1/2$, where sgn is the signum function.

The "strange attractor" looks like a strangely folded bow tie.

Julia Fractals

We now discuss some of the high points of this subject, studied worldwide by mathematicians and computer artists. There was a time when

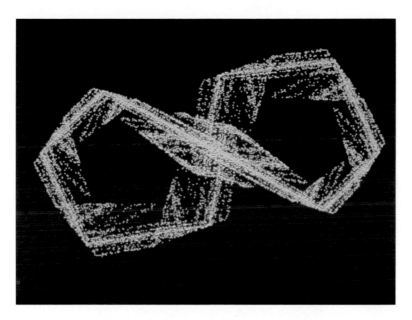

Figure 7.6 Fractal bow

they existed only in the mind of the French mathematician Gaston Julia. In 1919 he wrote a detailed "mémoire" of a few hundred pages that was awarded a prize by the French Academy. To present-day mathemati cians it is still an impressive study, and it is written in very readable French, but ... one will find hardly any pictures in it. The contents were ignored for about half a century, but now the book is right in the focus of attention. Computers have made visible that which could not be depicted in Julia's time. The visual results surpass all expectations. The recent book by Peitgen and Richter with the attractive title *The Beauty of Fractals* contains, apart from a number of mathematical analyses that can be understood only by specialists, a large number of colorful pictures of fractals and many examples of computer art. The most beautiful pictures have been made with powerful computers that exceed the budget of the private individual. However, making Julia fractals appear on the screen of a microcomputer is not too difficult. Here one has to be satisfied with less detail and possibly do without color.

Julia fractal derived from a simple function

Julia's work concerns iterative mappings of the type (C), but now the mapping $x \rightarrow f(x,y)$, $y \rightarrow g(x,y)$ has the property that all angles stay the same. So a square changes into another square, the sides of which *can* be curved. Figure 7.7 shows the effect this may have on a checkerboard. The mapping used is

(I)
$$x' = x^2 - y^2$$
$$y' = 2xy.$$

The checkerboard region is given by $1 \leq x \leq 2$, $1 \leq y \leq 2$. Mathematicians can deduce that the straight lines of the original checkerboard have changed into parts of parabolas.

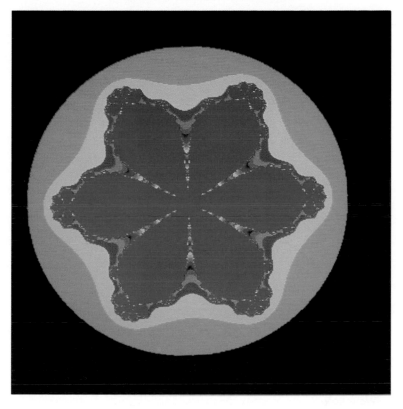

Julia fractal derived from a simple function

A mapping like (I) that leaves angles unchanged is called *conformal*. Conformal mappings occur quite a lot in mathematics and its applications. They are used in models of fluid flow, among other things. They also produce profiles of wings of airplanes. The theory behind them is based on the use of complex numbers and complex functions. As the theory of complex numbers is not included in the school curriculum, we will say no more on the subject in this chapter. In Appendix A, however, we will show how useful complex numbers can be in computer graphics, and in fractals, too. The somewhat more general mapping (A) expressed in iterative form is also conformal:

$$\text{(J)} \qquad \begin{aligned} x_{n+1} &= x_n^2 - y_n^2 + a \\ y_{n+1} &= 2x_n y_n + b. \end{aligned}$$

Here a and b are arbitrary constants.

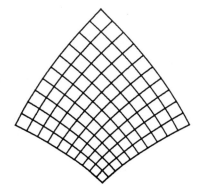

Figure 7.7 Conformal mapping of a checkerboard

THE JULIA SET. For the moment we concentrate on this model. We will soon realize that a whole world of fractals is concealed in it.

We follow Julia and study the orbit of an arbitrary starting point P_0, (x_0, y_0). We see that in a number of cases the orbit runs off to infinity. This is certainly the case if the distance between P_0 and the origin O is large enough. It can be verified that certainly all points P_0 for which

$$OP_0 > 1 + (a^2 + b^2)^{\frac{1}{4}}$$

produce such an orbit; we say they are attracted by infinity. We can divide the points P_0 in the plane into points whose orbit is attracted by infinity, and points for which this is not the case. The first collection of points constitutes the so-called area of attraction of infinity. The boundary of this area, a kind of watershed, is nowadays called a Julia set. It is nearly always a fractal!

Julia established a large number of properties of this set; we will mention only the most important ones. For convenience we will refer to the Julia set of (A) by J, or by $J(a,b)$ if the values of the parameters have to be given.

The first property of J is that the conformal mapping $x \rightarrow x^2 - y^2 + a$, $y \rightarrow 2xy + b$ transforms J into itself—in other words, leaves it invariant.

The second property of J is that if an arbitrary orbit has one point on J, then all its points will be on J. In other words, if P is a point of J, then all preceding and subsequent points are in J too. For computing purposes it is invaluable that all predecessors of a point cover J completely. A property of the conformal mapping (A) is that a given point has two predecessors. To see this we interpret (A) ($x' = x^2 - y^2 + a$, $y' = 2xy + b$) as two equations in which x' and y' are given, and from which x and y have to be derived. This is done as follows:

$$x^2 - y^2 = x' - a$$
$$(x^2 + y^2)^2 = (x^2 - y^2)^2 + 4x^2y^2 = (x' - a)^2 + (y' - b)^2$$

so that

$$x^2 + y^2 = \sqrt{(x' - a)^2 + (y\prime - b)^2}.$$

As we know both sum and difference of x^2 and y^2, we know each one of them individually; for instance:

$$x^2 = \frac{1}{2}(x' - a) + \frac{1}{2}\sqrt{(x' - a)^2 + (y' - b)^2}.$$

From this, two values of x follow, a positive one and a negative one. When x is known, y can be derived from $y = (y' - b)/(2x)$.

So we find that the mapping (A) can be inverted in two ways. Mathematicians say that this transformation has a two-valued inverse. Applied to the iterative system (J) this means that a point P_0 has two predecessors, which in turn have four, and so on. Once again we get the binary tree structure we know so well: 1, 2, 4, 8, 16, ... Thus as we go back, a point of J generates twice as many predecessors at each step. According to Julia they will distribute themselves over J in such a way that J is densely filled everywhere.

Another property of J is that all unstable periodic cycles are situated on J. An unstable periodic equilibrium repels points in its vicinity.

J is full of such points; consequently it is a *repeller*. An orbit of a point P_0 that does not lie on J will remove itself further and further away from it. An orbit of a point P_0 on J will stay on J. It can be either periodic or chaotic.

If, on the other hand, we look at the backward tracing iterative process for (J), in which x_n and y_n are derived from x_{n+1} and y_{n+1} in two ways by the calculation we have just given, J will be an attractor instead of a repeller. In this case all orbits either move toward J or stay on it in a stable way.

JULIA FRACTALS. Armed with our knowledge of fractals with a binary structure which we gained in previous chapters, it is very easy to make a computer program. We start with a point P_0 which we are sure is on J. We follow the orbit that goes backward from it, either with the Monte Carlo method or systematically with the backtrack method.

The latter was used in the program JULIAB. In this the calculation given above has been used to find the two predecessors of any given point. The fractal $J(0, 1)$ of Figure 7.8, a so-called dendrite, was generated like that. As initial point we chose $x_0 = -1.3002$, $y_0 = 0.6248$, an unstable fixed point.

Figure 7.9 gives us a picture of the fractal $J(-3/4, 0)$. Mandelbrot was one of the first to see this picture. Originally he used the backward iterative process just described, for which the Julia fractal is an attracting set. The picture reminded him of the famous church of San Marco in Venice, mirrored in the flooded San Marco square. The Julia fractals of (A) are always point-symmetric with respect to the origin. If, as here, $b = 0$, then they are also symmetric with respect to both the X-axis and the Y-axis. We can use this simple observation to speed up programs, doing the whole job in a quarter of the time, for instance. It is worth taking a closer look at the special case $b = 0$.

If in (A) we take $b = 0$ and if we start from a point $P\ (x, y)$ on the X-axis (in other words, a point for which $y = 0$), then it follows that the image point $P'\ (x', y')$ lies on the X-axis as well, because then $y' = 0$ too. For the corresponding iterative mapping (J), this means that the orbit of a point on the X-axis remains there. Consequently such an orbit (J) can be reduced to

(K1) $x_{n+1} = x_n^2 + a.$

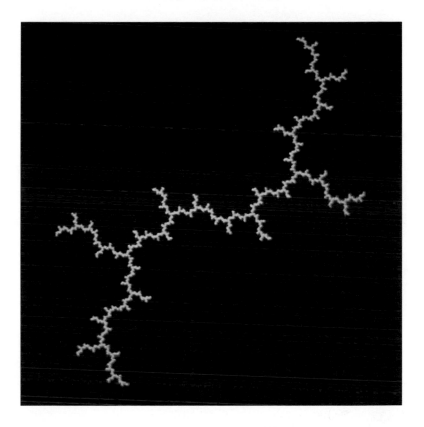

Figure 7.8 Julia fractal $J(0, 1)$, a dendrite

And now for the surprise. This iterative mapping is equivalent to the model of restricted growth we met in the previous chapter:

(K2) $\xi_{n+1} = \alpha \xi_n (1 - \xi_n)$

with a slightly different notation. To recognize the equivalence we just have to make the substitution $\xi_n = 1/2 - x_n/\alpha$. From (K2), (K1) then follows with $a = \alpha/2 - \alpha^2/4$.

In the previous chapter we found that for $1 < \alpha < 3$ (K2) has a stable fixed point $\xi = 1 - 1/\alpha$ and that after $\alpha = 3$ Feigenbaum's period doubling occurs. The value $\alpha = 3$ corresponds to $a = -3/4$,

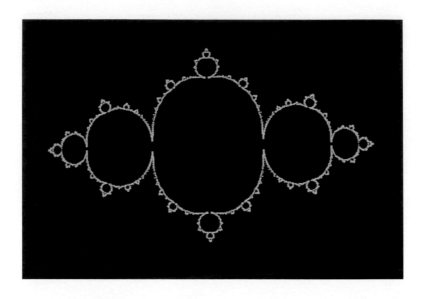

Figure 7.9 Julia fractal $J(-3/4, 0)$, the San Marco fractal

and the fixed point $\xi = 2/3$ to $x = -1/2$. So that fixed point is on the verge of being stable. For the plane mapping (J) this means that the fixed point $(-1/2, 0)$ is on the border between stability and instability and that according to Julia's theory it is part of the Julia fractal.

A further analysis shows that the San Marco fractal consists of an infinite series of islands that touch one another in pairs on the X-axis.

Figure 7.10 depicts the fractal $J(.11, .66)$. It is striking that in this case the figure is no longer connected; it consists of discrete points, just like Cantor's point-set. A fractal of this kind is usually called "Fatou" dust, in honor of the mathematician Fatou, a contemporary of Julia's.

The Julia fractals of (A) spark off many beautiful pictures, especially if one has the chance to use color. A number of these are shown in Peitgen and Richter's book. In this book we have also included a few such color illustrations.

SPECIAL CASE. Mathematicians are very interested in this kind of fractal. In order to understand something of their methods, we will

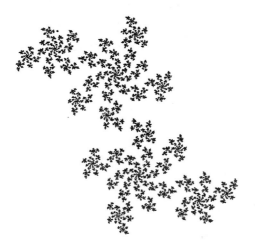

Figure 7.10 Julia fractal $J(.11, .66)$, scattered leaves

look at the special case $a = b = 0$, i.e., (I). We replace this scheme by a corresponding model in polar coordinates r, θ, in which $x = r \cos \theta$ and $y = r \sin \theta$. A small calculation shows that the model (J) can then be expressed as

$$r_{n+1} = r_n^2, \quad \theta_{n+1} = 2\theta_n \pmod{2\pi}$$

The nature of the orbit of a point P_0 with polar coordinates r_0, θ_0 depends on r_0 only. If $r_0 < 1$, in other words if P_0 lies within the unit circle, or to put it differently again if the distance of P_0 to the origin O is less than 1, then the orbit will spiral toward the origin. If $r_0 > 1$, the orbit will spiral to infinity. If $r_0 = 1$, the orbit will keep drifting around the unit circle forever. In this very special case the Julia fractal is identical to this circle. The distinctive appearance of a fractal has disappeared. The self-similarity of the circle, on the other hand, remains.

In order to gain some insight into an orbit on the unit circle, we write $\theta_0/(2\pi)$ as an indefinitely continuing binary fraction:

$$\theta_0/(2\pi) = 0.b_0 b_1 b_2 b_3 \ldots$$

Here the digits are either 0 or 1. We also agree to take the initial angle θ_0 between 0 and 2π. All angles derived from this we reduce to that same interval by subtracting 2π from them if necessary. Let us explain this with the aid of the following example. If $\theta_0 = 2\pi/7$, then $\theta_1 = 4\pi/7$, $\theta_2 = 8\pi/7$, $\theta_3 = 16\pi/7 = 2\pi/7$, so that $\theta_3 = \theta_0$. Here the orbit returns after three steps, so that we have a 3-cycle. Constructing a few more of these periodic orbits is simple. We can easily construct, for example (leaving out the factor $2\pi/31$),

1	2	4	8	16
3	6	12	24	17
5	10	20	9	18
7	14	28	25	19
11	22	13	26	21
15	30	29	27	23

This system immediately generates six 5-cycles, which are of course unstable. We can continue this way indefinitely. Thus mathematicians have found that the unit circle is filled entirely with periodic orbits— denumerably many, in fact. Still, an indenumerable number of all kinds of chaotic nonperiodic orbits remain.

What we have said concerning this special case is true for any Julia fractal. In Appendix A we demonstrate how we can get many more beautiful Julia fractals quite easily using complex numbers.

Mandelbrot

We can divide the Julia fractals into two main types. They can be either wholly disconnected or wholly connected. In the former case the fractal consists of indenumerably many discrete points. The classic example is Cantor's point-set spread over a line-segment.

If it is connected, on the other hand, the fractal consists of a succession of lines: sometimes a single closed curve; sometimes loops within loops within loops ... ; from time to time a dendrite.

With the Julia fractals $J(a,b)$ of the model (A), the type depends on the numbers a and b. Mandelbrot hit upon the fruitful idea of making a graph of this by interpreting a and b as the coordinates of a point and drawing that point if $J(a,b)$ is connected. At first glance this seems a rather complicated problem, but Julia had devised a trick for finding out whether or not $J(a,b)$ is connected without actually constructing $J(a,b)$ itself. One has only to examine the orbit of the starting point $x_0 = a$, $y_0 = b$ under the iterative process (J). If *this* orbit goes to infinity, then $J(a,b)$ is disconnected, like dust. With a simple computer program we can thus find out to which class $J(a,b)$ belongs. All points for which J is connected constitute the so-called Mandelbrot set. The result is Figure 7.11, an extremely peculiar figure that reminds one of a conglomerate of fruit; this is the so-called "potato man." Mathematically this structure is very simple. In Appendix A, which deals with complex numbers, we will go into it a bit more deeply. The part shaped like a kidney is bordered by a cardioid or heart-shaped curve. We express this mathematically by

$$4a = 2\cos t - \cos 2t$$
$$4b = 2\sin t - \sin 2t$$

Here t runs from 0 to 2π.

The adjoining circular part is indeed a circle with center $(-1,0)$ and radius $1/4$. Around it lie a series of both small and minute circular areas. More detailed pictures show that this phenomenon keeps repeating itself everywhere on an increasingly diminishing scale. In brief, the figure Mandelbrot found is a fractal!

On the X-axis the Mandelbrot fractal shows the bifurcation behavior of (K1) or (K2) we saw in the previous chapter (Figure 6.9). We emphasized the phenomenon of Feigenbaum's period doubling with universal scaling. That phenomenon was characterized by the values $\alpha = 3$, 3.4495, 3.5441, 3.5644, 3.5688, ..., 3.5699. Here these correspond to $a = -3/4$, $a = -5/4$, and $a = -1.3681$, -1.3940, -1.3996, ..., -1.4012. These values are rather special. They mark the points of tangency between circular areas whose diameters have been reduced by Feigenbaum's constant. We find these same areas around larger areas, and so on.

We saw that near $\alpha = 3.83$ there is an area with a stable 3-cycle. In Mandelbrot's figure this corresponds to a kind of island at $a = -1.75$, $b = 0$. On taking a closer look this island turns out to be a "continent" in miniature. We also find islands like this in the vicinity of $a = -0.12$ and $b = \pm 0.74$. These in their turn correspond to a stable 3-cycle of Julia's model (J).

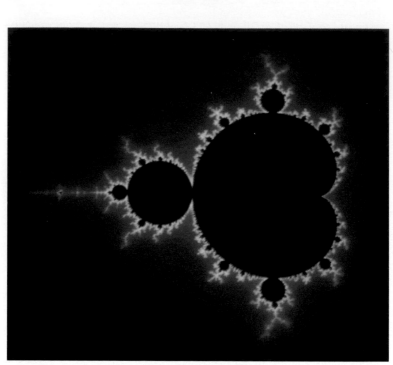

Figure 7.11 Mandelbrot fractal

When Mandelbrot saw this figure come into view on the screen for the first time and examined the hard copy more closely, he thought at first it was one single continent, a cardioid with adjoining circles. What he initially supposed to be bits of photographic dust turned out at successive enlargements to consist of miniature continents. Until recently it was indeed thought that the Mandelbrot fractal was built up of one large continent surrounded by an archipelago of denumerably many miniature continents. Very recently, however, a mathematician has been able to prove that everything is connected by meandering lines, like cobwebs! If we just relied on computer experiments this could never have been discovered, let alone proven. Ultimately man is superior to his machine!

ON THE MICROCOMPUTER. Good illustrations of the Mandelbrot fractal can generally be made only with faster computers, such as those with a 80286 or 80386 microprocessor, especially if one wants to use color. The use of TURBO BASIC saves a lot of time. However, a lot can be done with a microcomputer with a monochrome screen of 640×400 pixels, as the program MANDEL shows. In this program we center a rectangle on the screen. The number of vertical columns of pixels, $2n_1 + 1$, can be between 200 and 600, say. The number of horizontal pixel rows, $2n_2 + 1$, is determined by the size of the rectangle chosen and the pixel size. In this way a rectangle of $(2n_1 + 1)(2n_2 + 1)$ pixels is formed on the screen. Each pixel location corresponds to a pair of values (a, b) that have to be tested.

The test consists of a repeated application of the iterative process (J) with (a, b) as starting point. And the question is whether or not the orbit of (a, b) goes to infinity. If any point of the orbit lies outside the circle $x^2 + y^2 = 16$, we can be certain that the orbit goes to infinity, so we test for this. For an arbitrary starting point (a, b) we determine the number k of the iteration step at which a point of the orbit leaves the circle for the first time. Since this number might be very large—for points within the Mandelbrot fractal, of course, it is infinite—we limit the number of iteration steps, somewhat arbitrarily, to 50. Thus a k-value is associated with each pixel. On large computers this k-value can be taken as the definition of a color. (With an ordinary computer we just work modulo 2—in other words, white for even k, black for odd k [or vice versa]. In this way we would get Figure 7.11 with slightly less resolution, in black and white only.) We can produce really beautiful colored pictures by blowing up a small part of the Mandelbrot fractal. Figure 7.12 shows us the spectacular results that can be achieved with a better computer.

Figure 7.12 Detail of the Mandelbrot fractal

Julia fractal of the sine function

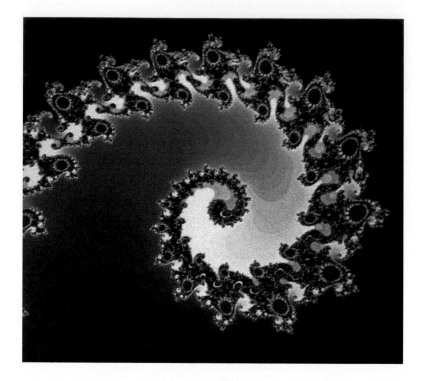

Detail of the Mandelbrot fractal

CHAPTER 8

Making Your Own Fractals

Constructing a fractal yourself is very enjoyable. It is an artistic activity that is closely related to the work of a visual artist or a graphic designer. We will not venture into making statements on art or computer art; opinions on this are too divided. It is tempting, however, to call some of the beautiful pictures that emerge on the screen of a computer "art." Anyone wanting to experiment in this direction is well advised to aim for a maximum of expression with a minimum of technical means. Colors may make the pictures more attractive, but one gets satisfactory pictures only if this is not at the expense of computing speed and screen resolution. That is why we restrict ourselves to a microcomputer with a high-resolution monochrome screen. A popular computer like the Olivetti M240 will do excellently. One then has 640×400 pixels, which is ideal. Computers with a resolution of 640×480 pixels are better still. It is advisable to use a compiled version of BASIC. The programs in this book can be executed in TURBO BASIC (now called POWER BASIC) without modifications except an adjustment of the screen number in the screen statement.

This chapter starts with a section on some technical matters of programming that help explain the computer programs we have used. These programs, listed in Appendix B, are written in TURBO BASIC. We have limited ourselves to those statements that are also used in other versions of BASIC. After that we will discuss a few types of fractals of various forms for which the program is either very simple or offers many possibilities. The idea is for the reader to get to work on these programs and introduce variations, either helped by the instructions given here or using his own imagination. The main programming principles are the Monte Carlo method discussed in Chapter 6 (used in the program DUST), the backtrack method discussed in Chapter 5 (the programs DUSTB and JULIAB), and the pixel method (the program MANDEL).

Programming

In this section we talk about the programs listed in this book. This will not trouble anyone who has a computer with the MS-DOS operating system. All the programs are written in TURBO BASIC (now called POWER BASIC) for a computer with a screen of 640×400 pixels. They can be run on an Olivetti M240 without any alterations. With other computers one will have to make minor adjustments, especially regarding graphics statements.

In the programs we have tried to be brief and clear. Comments are kept to a minimum, but obviously the reader can add his own comments after a "rem" statement. As a rule, variables and constants are given a single letter, followed by a second letter or number if necessary. The programs are identified by a mnemonic name of up to eight characters.

At the beginning of each program, the highest graphics resolution should be selected. We do this by the statement "screen 3," which means that we have 640×400 addressable pixels at our disposal. On some computers this statement has to be changed, by putting a different number after the word "screen," for instance. The next statement, "cls" (clear screen), ensures that we start with an empty screen. Both statements can be put on the same line, provided they are separated by a colon. This symbol enables one to put several small statements on the same line.

The number $\pi = 3.141593\ldots$ is not standardly available in TURBO BASIC and is defined as "pi" at the beginning of many programs.

To describe location on the screen, we use either so-called pixel coordinates or user-defined Cartesian coordinates. In pixel coordinates the statement

    pset(p,q)

means that a point is placed at (p,q), i.e., the pth position from left to right in the qth row from top to bottom. The positioning of a point comes down to the lighting up of a pixel, in green or amber, on a dark screen. The horizontal coordinate p is a whole number from 0 to 639, while the vertical coordinate q runs from 0 to 399. If we require pictures of the highest quality, pixel coordinates are preferable.

More often, we define a coordinate system adapted to the problem. For this we use the statement

window (x1,y1) — (x2,y2)

This statement means that the rectangular screen is described by

$$x_1 < x < x_2, \quad y_1 < y < y_2$$

The point (x_1, y_1) is the bottom left-hand corner and (x_2, y_2) the top right-hand one. If a "pset" statement is preceded by a "window" statement, the coordinates (p, q) are interpreted as the Cartesian coordinates defined by it.

To ensure that what is programmed as a square really is a square on the screen, we must take the rectangular form of the screen into account when using the "window" statement. On the Olivetti M240 this ratio is 4:3. A good example of this is the statement

window $(-4, -3) - (4,3)$

or

window $(-1.1, -.8) - (1.1,.8)$

If, on the other hand, we use pixel coordinates, we are faced with a further complication, as the pixel density is different in the horizontal and the vertical directions. Instead of a ratio of 4:3, we now have to work with the ratio 6:5.

In a few programs the positioning of a point depends on whether the outcome of a small calculation is 0 or 1. The appropriate statement is then

pset(p,q),k

where $k = 0$ or 1. If $k = 1$, a point is placed at (p, q), i.e., a pixel lights up. If $k = 0$, then a black point is placed on a black background. Either nothing happens, or a point already placed there disappears; in other words, the pixel goes out.

With the statement

line(p1,q1) — (p2,q2)

in TURBO BASIC, a straight line is drawn from (p_1, q_1) to (p_2, q_2). As with the "pset" statement, the coordinates are either pixel coordinates or, after a "window" statement, Cartesian coordinates. When we want to link up a succession of line-segments, the starting position can be omitted if it is also the end position of the previous "line" statement. In that case we use the version

line — (p,q)

Here p and q are the absolute coordinates of the endpoint of the line-segment drawn. Just as in the "pset" statement, the "line" statement can be followed by ",k" with the same meaning.

In TURBO BASIC a circle can be drawn with the special statement

circle (p,q),r

Here (p, q) is the center and r is the radius.

There are various ways of making small changes in a program. A number of programs contain parameters a, b, c, ..., for which different values can be chosen. If they are defined in a certain program line, that line has to be changed to get a different choice of numbers. The "edit" command of MS-DOS makes this very easy. It is also possible to introduce parameter values via the keyboard. To do this the program contains an "input" statement.

Dust Clouds

Fractal structures built up from points are a nice subject for experiment. The programs are extremely simple, and endless variations are possible. Examples are provided by the programs MIRA and DUST we discussed in the previous chapter.

The program MIRA is straightforward, and it is easy to make variations on it. With the program DUST we encounter a tree structure. Here the Monte Carlo method presents relatively few programming problems, and the resulting pictures are drawn very quickly. A better and more systematic method, however, is the backtrack method discussed in Chapter 5.

The basis of the first type of program is a mathematical rule by which from any point P a new point $F(P)$ can be derived. Geometrically this represents an iterative mapping or a repeated transformation. An arbitrary starting point P_0 then leads to a point-series P_0, P_1, P_2, P_3, ..., the orbit of P_0. Beautiful pictures can generally be made from one or more orbits of well-chosen starting points.

MIRA AND ITS VARIATIONS. We now focus our attention on the program MIRA and possible variations. As we said in Chapter 7, this is all based on an iterated series x_0, x_1, x_2, x_3, ... Here, for all values of n,

$$x_{n+1} = (1 + b)F(x_n) - bx_{n-1}$$

where

$$F(x) = ax + (1 - a)\frac{2x^2}{1 + x^2}.$$

We can derive the following iterative mapping from this series:

(A)
$$x_{n+1} = by_n + F(x_n)$$
$$y_{n+1} = -x_n + F(x_{n+1})$$

In the program this is incorporated efficiently as a sequence of four statements:

z=x : x=by+w · w=F(x) : y=w-z

Our standard choice is

(B) $$F(x) = ax + (1 - a)\frac{2x^2}{1 + x^2}.$$

We now have two parameters to be chosen at will. We get the most interesting pictures by choosing $-1 < a < 1$ and taking $b = 1$ or a little smaller, for instance $b = 0.99$. Every choice of a and b offers a potentially interesting picture.

To begin with we restrict ourselves to $b = 1$. In that case the iterative mapping is area-conserving, so we can get various types of orbits: periodic cycles, closed curves, island structures, and finally chaos.

The possibility of composing a picture from different types of orbits is very attractive. The choice can be made on artistic grounds. A wide field of computer art is open here!

In the table below we give a number of possibilities. In a few cases something beautiful, a so-called aesthetic chaos, can be obtained by choosing a single chaotic orbit.

Program MIRA with $b = 1$

a	Window	Starting points
-0.4	$(-24, -18)$-$(24, 18)$	$4, 0$
-0.4	$(-40, -32)$-$(40, 32)$	$20, 0$
-0.2	$(-24, -18)$-$(24, 18)$	$10, 0$
0.3	$(-6, -4.5)$-$(6, 4.5)$	$2, 0$
0.31	$(-60, -45)$-$(60, 45)$	$12, 0$
0.32	$(-20, -15)$-$(20, 15)$	$-5, 0$
0.30	$(-40, -30)$-$(40, 30)$	$7, 0$ $-12, 0$ $-21, 0$
-0.05	$(-40, -30)$-$(40, 30)$	$2, 0$ $7.5, 0$ $9.8, 0$ $15, 0$ $18, 0$ $20, 0$ $25, 0$
0.18	$(-32, -24)$-$(32, 24)$	$5.3, 0$ $8, 0$ $9, 0$ $15, 0$ $20, 0$

Program MIRA with $b < 1$

a	b	Window	Starting point
-0.48	0.93	$(-16, -14)$-$(16, 10)$	$4, 0$
-0.4	0.99	$(-16, -12)$-$(16, 12)$	$4, 0$
-0.25	0.998	$(-16, -12)$-$(16, 12)$	$3, 0$
0.01	0.96	$(-16, -12)$-$(16, 12)$	$3, 0$
0.1	0.99	$(-12, -9)$-$(12, 9)$	$3, 0$
0.7	0.9998	$(-20, -15)$-$(20, 15)$	$0, 12$

If a is close to 1, the orbits are generally fairly regular, but as we move away from this value their behavior will become more irregular

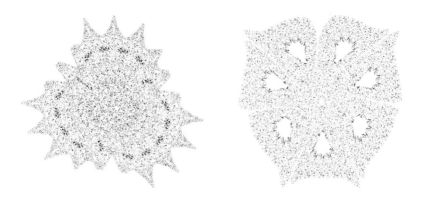

Aesthetic chaos with MIRA

Feather-shaped chaos with MIRA

and chaotic. If we take b slightly smaller than 1, this means geometrically that successive iterations reduce areas by the factor b. This gives the orbits a tendency to spiral inward and converge to a kind of limit structure, which may have an aesthetically pleasing shape.

It is usually sufficient to generate a single orbit with a well-chosen starting point. In fact, in this case the choice does not really matter, as all orbits have the same limit structure in the end. In the program we can start with an arbitrary point and then leave out the first couple of hundred points of the orbit. The number of points to be drawn depends on computing speed and the resolution of the screen. We normally draw a few thousand points per orbit.

In the model (A) one can choose all kinds of functions for $F(x)$. Anyone who wants to have a go at this himself must take into account the general rule that for large values of x, $F(x)$ should not increase faster than a linear function. The following functions are examples that satisfy this:

$$F(x) = ax + c\sin x$$
$$F(x) = ax + c\cos x$$
$$F(x) = a + c\sin x$$
$$F(x) = ax + \frac{cx^2}{1 + |x|}$$

Functions with a multiple function prescription are also possible, such as

$$
\begin{aligned}
F(x) &= ax & &\text{for } |x| < 1 \\
F(x) &= ax + c(x - 1) & &\text{for } x > 1 \\
F(x) &= ax + c(x + 1) & &\text{for } x < -1
\end{aligned}
$$

In this example the graph of $F(x)$ is composed of straight lines like an elongated letter Z. With this choice we still have some idea what kind of picture to expect. If $|a + c| < 1$, the iterative mapping will behave roughly like a repeated rotation through the angle α for which $\cos \alpha = a + c$. If $a + c = 1/2$, say, $\alpha = \pi/3$; a rotation of a sixth of a turn results.

For $b = 1$ we get periodic orbits, island structures, and chaos. Of the many combinations possible with $b = 1$, we mention

1. $a = 2$, $c = -2$, $\alpha = \pi/2$ with a chaotic orbit generated from $x = 0.1$, $y = 0$. One needs to take as many as $50\,000$ points of this.

2. $a = 3.5$, $c = -3$, $\alpha = \pi/3$ with a chaotic orbit from $x = 0.1, y = 0$.

3. $a = 0.9$, $c = -1.4$, $\alpha = 2\pi/3$ with normal orbits from (4.8,0) and (12,0) and chaotic orbits from (2,0) and (11,0).

A number of the possibilities we have just mentioned are incorporated in the programs CLOUD, CLOUD1, and CLOUD2.

DUST PROGRAMS. A program for tree-structure fractals, such as DUST or DUSTB, is just as flexible as the program MIRA we discussed just now. If we restrict ourselves to a binary tree structure, there are now two transformations **L** and **R** by which two new points can be generated from a given point. So a starting point has a continually doubling number of successors: 1, 2, 4, 8, 16, ..., so that it is better to speak of a dust cloud than of an orbit. The actual fractal is the limit set of all these points. If the starting point is part of the fractal, which is the case if it is a fixed point of **L** or **R**, then all points derived from it will be a part of it too. Suppose we go up to the order 10; then we have $1 + 2 + 4 + 8 + \cdots + 1024 = 2047$ points already. Given the resolution of the screen, this usually gives a good enough approximation. The backtrack method is the most efficient for generating all these points.

All this has been incorporated in the program DUSTB. In Chapter 5 we discussed this method. In DUSTB both **L** and **R** are contracting rotations:

$$(C) \quad \mathbf{L} \begin{cases} x_{n+1} = ax_n - by_n \\ y_{n+1} = bx_n + ay_n \end{cases}$$

and

$$(D) \quad \mathbf{R} \begin{cases} x_{n+1} = cx_n - dy_n + 1 - c \\ y_{n+1} = dx_n + cy_n - d \end{cases}$$

L is a rotation-enlargement with the origin as center and reduction factor $\sqrt{a^2 + b^2}$. **R** is a rotation-enlargement about $(1,0)$ with reduction factor $\sqrt{c^2 + d^2}$.

The Monte Carlo version of the program, DUSTV, is very effective in practice. Here a random choice between **L** and **R** is made at each iteration step. The program has roughly the same form as MIRA. The heart of it is

```
for k=1 to 10000
if rnd < .5 then gosub L else gosub R
pset (x,y)
next k
```

Using this method one can also exploit possible symmetries in the fractal by simultaneously drawing each point together with those points symmetrical to it. This is the method we followed when we wrote the program DUST of Figure 6.2.

Now the way is wide open for variations and generalizations. For the constants a, b, c, d of $\mathbf{L}$ and $\mathbf{R}$ we can try using all kinds of numbers, provided that $a^2 + b^2 \leq 1$ and $c^2 + d^2 \leq 1$. In Chapter 5 we showed that $\mathbf{L}$ and $\mathbf{R}$ can also be what we called contraction-mirrorings. More generally still, $\mathbf{L}$ and $\mathbf{R}$ can be arbitrary contracting transformations. One needs to make the origin a fixed point of $\mathbf{L}$ and $(1,0)$ a fixed point of $\mathbf{R}$. For $\mathbf{L}$ one can take a general linear transformation—say,

$$(E) \quad \begin{aligned} x_{n+1} &= a_1 x_n + a_2 y_n \\ y_{n+1} &= a_3 x_n + a_4 y_n \end{aligned}$$

The quantity $a_1 a_4 - a_2 a_3$ gives the area scale factor. With the mapping (E) a square is transformed into a parallelogram. Contraction requires $|a_1 a_4 - a_2 a_3| < 1$. The sign of $a_1 a_4 - a_2 a_3$ indicates whether or not the transformation involves a reflection.

This sort of thing is very familiar to mathematicians. For those unacquainted with the mathematical significance of these remarks, experimenting on the computer is the obvious way to familiarize oneself with the properties of linear transformations. For $\mathbf{R}$ one can take something similar. The program allows a number of possibilities. On the following page we give just one example of a beautiful picture.

1. Both $\mathbf{L}$ and $\mathbf{R}$ are rotations like (C) and (D). Take, for instance,

a	b	c	d
$\frac{1}{2}$	$\frac{1}{2}$	$\frac{1}{2}$	0
$\frac{1}{4}$	$\frac{1}{4}\sqrt{3}$	$\frac{1}{4}$	$-\frac{1}{4}\sqrt{3}$
$\frac{1}{4}$	$\frac{1}{4}\sqrt{3}$	$\frac{1}{4}$	$-\frac{1}{4}\sqrt{3}$
$-\frac{1}{4}$	$\frac{1}{4}\sqrt{3}$	0	$\sqrt{2}$

Try other values yourself, replacing $\mathbf{R}$ by a reflection, for instance.

At this point we would like to remind the reader that for a rotation $\mathbf{L}$ with angle of rotation ϕ and reduction factor r, the parameters a and b are given by $a = r \cos \phi$, $b = r \sin \phi$.

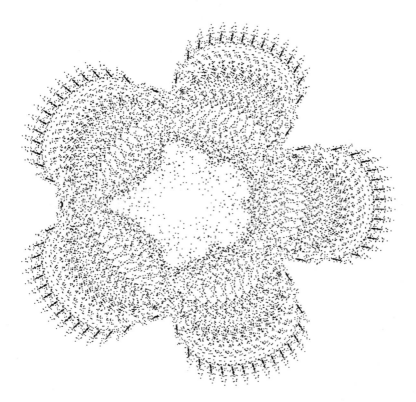

Flower-shaped chaos with MIRA

2. For **R** we can take a quadratic transformation such as

$$x' = (x^2 - y^2 + 1)/2, \quad y' = xy$$

Combined with the contraction-mirroring for **L**

$$x' = \tfrac{1}{2}x + y, \quad y' = -x - \tfrac{1}{2}y$$

the resulting picture is very interesting. By choosing **L** to be

$$x' = \tfrac{1}{2}y, \quad y' = x + y$$

success is also guaranteed.

Meanders

Fractals that consist of meandering lines also lend themselves to computer experiments. In such an experiment we make an approximation to a fractal by means of a sequence of line-segments. The fractals of von Koch (Figure 3.1) and Minkowski (Figure 3.9) are examples of such fractals.

As we saw in Chapter 3, the construction of these fractals can be based on a number system. We call the base of this number system ν, and the order of the approximation we want, p. The number of line-segments will then be $\nu \uparrow p$ or a number proportional to this. To number the line-segments we use the index n. As a rule both the length and the direction of the nth line-segment are determined by the expansion of n in the base-ν number system.

In the case of von Koch's, Minkowski's, and Lévy's (Figure 3.19) fractals, all line-segments are equally long, and only their direction depends on n. For fractals produced by the program MEANDER both the length and the direction of the nth line-segment may depend on the index n. For the fractals produced by STAR the direction follows a simple rule: each line-segment has been rotated by a constant angle relative to the previous line-segment. These fractals are very suitable for experimentation. So we will now add to what we said on this subject at the end of Chapter 4.

ADDITIONAL REMARKS. We begin by recalling the general construction. The index n runs from 0 to $(\nu + 1)\nu \uparrow (p - 1) - 1$. We call the constant angle of rotation α, so that the direction of the nth line-segment is $n\alpha$. The length is determined by the number of factors ν in p. If that number is $p - 1$ or p, we get the largest length, 1, say. If that number is $p - 1 - k$, with $k = 0, 1, \ldots, p - 1$, then the length is $r \uparrow k$. Here r is a reduction factor (between 1 and 0) selected beforehand.

In principle, every choice of p, ν, α, and r will result in a meandering line. If we want the meandering line to close on itself, in order to possibly form a beautiful symmetrical figure, then α has to be chosen in such a way that $(\nu + 1)\alpha$ is a multiple of 2π. The star fractal of Figure 4.24 corresponds to $p = 5$, $\nu = 4$, $\alpha = 4\pi/5$, $r = 0.35$.

Anyone who wants to experiment does best by first selecting ν and α. Then one starts with a low order p: 2 or 3, for instance. One

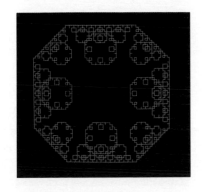

Embroidery pattern of two Lévy fractals

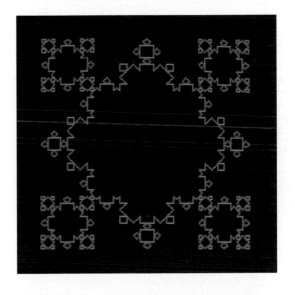

Result of an experiment with a meander program

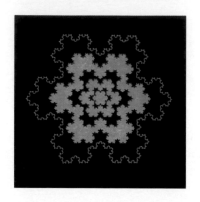

Fractal formed from a series of Koch islands

fixes the reduction factor r experimentally, depending on the aesthetic quality of the picture obtained. After that one can take p as high as computing speed and resolution allow.

The table below shows some possibilities. Since it is very difficult to predict the size of the entire figure in the various cases, screen co-ordinates have to be introduced in practice with a "window" statement.

ν	α	r	p
2	$2\pi/3$	0.62	9
3	$\pi/2$	0.47	6
4	$2\pi/5$	0.5	3
4	$4\pi/5$	0.35	4
5	$\pi/3$	0.33	5
6	$2\pi/7$	0.383	4
6	$6\pi/7$	0.32	4
7	$\pi/4$	0.5	4
8	$2\pi/9$	0.25	4
11	$\pi/6$	0.5	3
14	$14\pi/15$	0.3	3
15	$\pi/8$	0.27	3
19	$\pi/10$	0.5	2

REPEATED MOTIF. In Chapter 3 we focused attention on a group of fractals generated by repeating a simple motif. We started with a baseline or starting line—a square, say. At each cycle each line-segment was replaced by the motif, as in Figure 3.12, for instance. This program, MEANDER, offers many opportunities for variations, as we can see in Figures 3.13–3.20. It makes heavy use of memory space, which can lead to problems on some computers. For this reason we included two different programs (MEANDERN and MEANDERB) that do not have this drawback. In the first one, MEANDERN, we use the so-called number system method. We suppose, just as we did in Chapter 3, that the motif consists of ν vectors, and that with a chosen order p and a chosen baseline we have to draw $\nu\!\uparrow\!p$ line-segments. These we number from 0 to $\nu\!\uparrow\!p - 1$ and we determine the expansion of a given index n in the base-ν number system $n = t_{p-1}t_{p-2}\ldots t_1t_0$, i.e.,

$$n = t_0 + \nu t_1 + \nu 2 t_2 + \cdots + \nu^{p-1}t_{p-1}.$$

In this program the ν vectors of the motif have to be fixed by giving the length and direction (in degrees, say) of each vector. If we restrict ourselves, as we did in the program, to the motif of von Koch's fractal, the data are

$$\tfrac{1}{3},0, \qquad \tfrac{1}{3},60, \qquad \tfrac{1}{3},-60, \qquad \tfrac{1}{3},0,$$

or $l(0)$, $\phi(0)$, $l(1)$, $\phi(1)$, $\ldots$, $l(\nu - 1)$, $\phi(\nu - 1)$ in a general notation. The length l and direction ϕ of the nth vector are then given by

$$l = l(t_0)l(t_1)l(t_2)\ldots l(t_{p-1}),$$
$$\phi = \phi(t_0) + \phi(t_1) + \phi(t_2) + \cdots + \phi(t_{p-1}).$$

That is the heart of the program. The remainder is preparatory book-keeping and trimmings.

The data of the baseline are also introduced as a series of vectors. So the length and direction (in degrees) of each line-segment in the baseline are listed. For instance, with a square this would be

$$2,\ 0,\ 2,\ 90,\ 2,\ 180,\ 2,\ -90$$

In the program the degrees have to be converted into radians by $180° = \pi$ radians. We also have to mark the point from which the meander starts. If it starts in $(-1, -1)$, for instance, then the baseline is a square with vertices $(-1, -1), (1, -1), (1, 1), (-1, 1)$.

To help the user experiment, only the *relative* sizes of the lengths of the component vectors of the model have to be given.

So for von Koch's motif the data

$$1, \ 0, \ 1, \ 60, \ 1, \ -60, \ 1, \ 0$$

are sufficient.

One can see in the program how these relative lengths are converted to absolute ones.

When we discussed the meander fractals in Chapter 3, we demonstrated how one can experiment with this. We will now list the main suggestions for MEANDERN.

Baseline	Data	Starting point
line-segment	$2, 0$	$(-1, 0)$
double line-segment	$2, 0, 2, 180$	$(-1, 0)$
equilateral triangle	$3, 300, 3, 180, 3, 60$	$(0, \sqrt{3})$
equilateral triangle	$3, 240, 3, 0, 3, 120$	$(0, \sqrt{3})$
square	$2, 0, 2, 90, 2, 180, 2, -90$	$(-1, -1)$
square	$2, 90, 2, 0, 2, -90, 2, 180$	$(-1, -1)$

Motif type	ν	Data
Lévy	2	$1, 45, 1, -45$
Minkowski	3	$1, 45, 2, -45, 1, 45$
von Koch	4	$1, 0, 1, 60, 1, -60, 1, 0$
Koch square	5	$1, 0, 1, 90, 1, 0, 1, -90, 1, 0$

Introducing variations oneself is easy; there is a good chance the results will be unexpected and at times beautiful. We illustrate this with the following example. For a baseline we take an inward square. A possible motif with variations for $\nu = 6$ is

1, 0, 1, 90, 1, −45, 1, 45, 1, −90, 1, 0
1, 0, 1, 90, 2, −45, 2, 45, 1, −90, 1, 0
1, 0, 1, 90, 1, −60, 1, 60, 1, −90, 1, 0
1, 0, 1, 90, 2, 45, 2, −45, 1, −90, 1, 0

As we said earlier, the original program MEANDER of Chapter 3 is fastest. The program we are discussing now is a bit slower, if only because for each index its expansion in the base-ν system has to be determined. We have included a second version of this program, in which the backtrack method was used. This program is a bit more complicated, because we have to make it possible to use the backtrack method in any number system we like. For this reason all intermediate calculations are carried out in a double array $x(\nu,p)$, $y(\nu,p)$, ν being the number of elements of the model and p the order of approximation ($\nu = 6$ and $p = 3$, say, as in the previous example). This program, MEANDERB, is just as flexible as the previous one, but considerably faster.

Complex Numbers

In this appendix we first summarize the essentials of the theory of complex numbers for the reader who is already familiar with it to some degree, or has at least come across it before. Then we will demonstrate how useful complex numbers are in the description and analysis of some fractals, especially those of Julia and Mandelbrot.

In essence, calculating with complex numbers corresponds to geometrical computations in the plane. A complex multiplication, for instance, is the same as a rotation combined with a rescaling!

SUMMARY. Complex numbers are defined in terms of the so-called imaginary unit i (sometimes written as j) with the property

$$i^2 = -1.$$

A complex number looks like $a + ib$ or $a + bi$, a and b being real numbers. The number a is called the real part and b the imaginary part. The notation for this is

$$c = a + bi, \quad a = \operatorname{Re} c, \quad b = \operatorname{Im} c.$$

We imagine the complex number $a + bi$ represented in the plane by the point $P(a, b)$ in Cartesian coordinates (Figure A.1). The X-axis is often called the real axis, as it carries all real numbers. The Y-axis is called the imaginary axis.

The polar coordinates of $P(a, b)$, r and ϕ, are called the absolute value and the argument of the complex number $c = a + bi$, respectively. So:

$$r = |c| = \sqrt{a^2 + b^2}$$

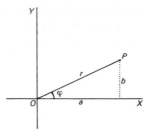

Figure A.1 Representation of a complex number in the complex plane

and

$$\phi = \arg c = \arctan(b/a) \quad (\pm\pi).$$

Note that the argument ϕ of c is taken to be between $-\pi$ and $+\pi$. We always have $\tan\phi = b/a$, but on the other hand $\phi = \arctan(b/a)$ only if $a > 0$. Since $a = r\cos\phi$ and $b = r\sin\phi$, we can also write the complex number $c = a + bi$ as

(1) $\quad a + bi = r(\cos\phi + i\sin\phi).$

The numbers $a + bi$ and $a - bi$ are called complex conjugates. The corresponding points in the plane are mirror images of one another with respect to the real axis. In abbreviated notation, we indicate the complex conjugate of c by $\bar{c}$.

With complex numbers we can carry out ordinary algebraic operations provided we keep in mind that $i^2 = -1$. For example:

$$(a + bi)(a - bi) = a^2 + b^2,$$
$$\frac{1}{a + bi} = \frac{a - bi}{a^2 + b^2},$$
$$(a + bi)^2 = (a^2 - b^2) + 2abi.$$

A complex number c for which $|c| = 1$ is represented by a point on the unit circle. For the product of two such numbers one has the beautiful equation

(2) $\quad (\cos\alpha + i\sin\alpha)(\cos\beta + i\sin\beta) = \cos(\alpha + \beta) + i\sin(\alpha + \beta)$

COMPLEX NUMBERS 175

which can be derived from well-known trigonometric formulas. So multiplication comes down to the adding of the arguments! More generally, for the product of two complex numbers c_1 and c_2 we have the rule

(3) $$|c_1 c_2| = |c_1| \cdot |c_2|,$$
$$\arg c_1 c_2 = \arg c_1 + \arg c_2.$$

RELATIONSHIP WITH TRANSFORMATIONS. Calculations with complex numbers correspond to geometrical transformations. A transformation of a complex number is expressed as

(4) $$z' = F(z).$$

Here $z = x + yi$ and $z' = x' + y'i$ in Cartesian coordinates. We now show how in a number of simple cases this corresponds to a geometrical transformation:

1. $z' = z + c$, a translation or parallel motion. In real notation this is the same as $x' = x + a$, $y' = y + b$.
2. $z' = iz$, a rotation by $\pi/2$ or a quarter turn. In real notation: $x' = -y$, $y' = x$.
3. $z' = (\cos \alpha + i \sin \alpha)z$, a rotation through the angle α. In real notation: $x' = x \cos \alpha - y \sin \alpha$, $y' = x \sin \alpha + y \cos \alpha$.
4. $z' = az$, where a is real, a *central enlargement* or rescaling by a factor a.
5. $z' = \bar{z}$, reflection or mirroring in the real axis.
6. $z' = (\cos \alpha + i \sin \alpha)\bar{z}$, a reflection in the line through the origin with the direction $\alpha/2$. In real notation: $x' = x \cos \alpha + y \sin \alpha, y' = x \sin \alpha - y \cos \alpha$.

Combining 3 and 5 gives us complex multiplication:

$$z' = cz.$$

Geometrically, this is a rotation with rescaling, a similarity transformation in which the origin is the center or fixed point. We define a general similarity transformation in which an arbitrary point, z_0, is the center, in terms of complex numbers by

$$z' = z_0 + c(z - z_0).$$

APPLICATION TO FRACTALS. In these terms it is now very simple to describe the structure of the fractals discussed in Chapter 5. These were built up from two similarity transformations **L** and **R**. In complex notation the description of the fractal of Figure 5.21 is

L $z' = \frac{1}{2} + \left(\frac{1}{6}i\sqrt{3}\right)\bar{z},$

R $z' = \frac{1}{3} + \frac{2}{3}\bar{z}.$

The fractal of Figure 5.22 is analogously described as

L $z' = cz,$

R $z' = c\bar{z} + (1 - c),$ with $c = \frac{3}{10}(-1 + i\sqrt{3}).$

COMPOUND TRANSFORMATIONS. We can also describe compound transformations like $\mathbf{R}^2$, **RLR**, and so on in complex numbers. In the example just given we have $c\bar{c} = \frac{9}{25}$, and $\mathbf{R}^2$ can be calculated as

$$z'' = c\bar{z}' + (1 - c) = c(\bar{c}z + 1 - \bar{c}) = c\bar{c}(z - 1) + c$$

or

$$z'' = \frac{9}{25}(z - 1) + c.$$

We recognize here a central enlargement. We find the center by putting $z'' = z$, so $z = (-25 + 15i\sqrt{3})/32$.

The combination $\cos\alpha + i\sin\alpha$ is often written as $\exp(i\alpha)$ or $\exp(\alpha i)$ with the notation of the exponential function. The usual rules for that function such as $\exp(0) = 1$, $\exp(a + b) = \exp(a) \cdot \exp(b)$ still apply. Sure enough, in this notation formula (2) can be written as

$$\exp(i\alpha) \cdot \exp(i\beta) = \exp\big(i(\alpha + \beta)\big).$$

A consequence of this notation is the emergence of the "mysterious formula" $\exp(i\pi) = -1$.

THE JULIA FRACTAL. Complex numbers are of special importance in the mathematical description of Julia fractals. These fractals are determined by endlessly repeating transformation (4). We have to be able to take the derivative of the function $F(z)$. For simple functions, however, like a polynomial, sine, cosine, and so on, this is done in the same way as for ordinary functions of real numbers. Thus the derivative or differential coefficient of the function $F(z) = z^m$ is given by $F'(z) = mz^{m-1}$. Just as for real functions, a linear approximation of a complex function $F(z)$ can be made using this in a small region of z-values. If we restrict ourselves to values close to z_0, the approximation

(5) $F(z) \approx F(z_0) + (z - z_0)F'(z_0)$

holds.

This is an important observation, as it means that for $z \approx z_0$ the mapping $z' = F(z)$ is equivalent to a combination of a translation, a rotation, and an enlargement (unless $F'(z_0) = 0$). Consequently this is a similarity transformation with scale factor $|F'(z_0)|$. We can imagine the transformation $z' = F(z)$ to be composed of similarity transformations with a scale factor that changes from point to point. Since angles do not change under a similarity transformation, the same holds for the transformation $z' = F(z)$. In Chapter 7 we used the term conformal mapping for these transformations, and we illustrated this by some examples. Our main one was

$$x' = x^2 - y^2 + a,$$
$$y' = 2xy + b.$$

At first sight this seems a rather artificial combination. It is only in complex notation that we see how logical and simple this really is:

(6) $z' = z^2 + c.$

Substituting $z = x + yi$ and $c = a + bi$ and splitting (6) into real and imaginary parts immediately leads to the desired result.

The fixed or equilibrium points and the periodic points of $z' = F(z)$ are especially important. If z is a fixed point, then it follows from the linear approximation (5) in the vicinity of the fixed point that stability is determined by the scale factor $|F'|(z)$. So if we look at the iterative process

(7) $z_{n+1} = F(z_n)$,

and if for the fixed point z $|F'(z)| < 1$, then the orbit of a point z_0 in the vicinity of z will run toward the equilibrium point—is attracted by it, as it were. In this case the equilibrium is stable or attracting. If on the other hand $|F'(z)| > 1$, then the orbit is repelled by z_0; the equilibrium is unstable or repelling. If finally $|F'(z)| = 1$, then we cannot say anything about the stability of the orbit. The equilibrium is now called neutral.

For a periodic cycle, stability can be examined in a similar way. The mapping (6) has two fixed or equilibrium points. They can be calculated by taking $z' = z$ and then solving the quadratic equation

$$z^2 - z + c = 0.$$

If we call the roots of this equation ζ_1 and ζ_2, then $\zeta_1 + \zeta_2 = 1$. Geometrically this means that the midpoint of the points ζ_1 and ζ_2 is the point representing $1/2$ (Figure A.2). From (7) with $F(z) = z^2 + c$ and $F'(z) = 2z$ it follows that the stability of the equilibria depends on $|2z|$. In Figure A.2 we can see that for one of the two equilibria, ζ_2 say, $|\zeta_2| > 1/2$ always, so this equilibrium is unstable. The other equilibrium, ζ_1, can be either stable or unstable. A small calculation gives $\zeta_1 = 1/2 - (1/4 - c)^{1/2}$. Geometrically the condition of stability $|\zeta_1| < 1/2$ means that ζ_1 lies within a circle around the origin O with radius $1/2$.

Calculating the 2-cycle of (6) is not difficult either. If the mapping (6) transforms z into w and w back into z, then

$$w = z^2 + c, \qquad z = w^2 + c.$$

A simple calculation shows that w and z satisfy

$$w + z = -1, \qquad wz = c + 1.$$

It is possible to show that the condition for stability of the 2-cycle is

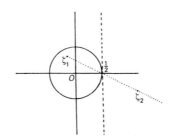

Figure A.2 Positions of the equilibrium points for the quadratic mapping

$|wz| < 1/4$. This means $|c + 1| < 1/4$, or geometrically that c lies within the circle with center -1 and radius $1/4$. This result will be important in the analysis of the Mandelbrot fractal.

THE MANDELBROT FRACTAL. In Chapter 7 we said something about the Mandelbrot fractal without using complex numbers. With complex numbers it is all much more logical and easy. We remind the reader that the Mandelbrot set of $z' = z^2 + c$ consists of the points c for which the orbit of $z = 0$, i.e.,

$$0, \; c, \; c^2 + c, \; (c^2 + c)^2 + c, \; \ldots$$

stays bounded. We also recall a most remarkable property of the Mandelbrot set: it contains a kidney-shaped area with a circle attached to it. It is relatively easy to explain why this is the case. If there is a stable fixed point, then the orbit of $z = 0$ will run toward it. This happens for $|2\zeta_1| < 1$ where $\zeta_1 = 1/2 - \sqrt{1/4 - c}$, as we saw just now. At the margin of the area of stability $|2\zeta_1| = 1$ holds, so that we can take $2\zeta_1 = \cos\theta + i\sin\theta$. Substituting in $2\zeta_1 = 1 - \sqrt{1 - 4c}$ gives, after rearranging,

$$\operatorname{Re} c = \tfrac{1}{4} + \tfrac{1}{2}\cos\theta(1 - \cos\theta),$$
$$\operatorname{Im} c = \tfrac{1}{2}\sin\theta(1 - \cos\theta).$$

If we let θ run from 0 to 2π, then the curve defined by this turns out to be exactly the cardioid by which the kidney-shaped area is bounded. The circular area nearby corresponds with a stable 2-cycle. We have just deduced that the 2-cycle is stable for c-values that satisfy $|c + 1| < 1/4$. For these values the orbit of $z = 0$ is bounded, so again we get a part of the Mandelbrot set. In this case this is the disk $|c + 1| < 1/4$ around $c = -1$ with radius $1/4$.

APPENDIX B

Programs

Chapter 1

TREEH1 H-fractal
TREEH2 H-fractal, backtrack method
TREE2 Binary tree
TREE3 Ternary tree
SIER Sierpinski sieve

Chapter 2

COMB Cantor comb

Chapter 3

KOCH Koch curve
MINK Minkowski sausage
MEANDER Island with meandering coastline
LEVY Lévy curve
DRAGON Dragon curve with arbitrary angle
DRAGON0 Dragon curve between two points
DRAGON1 Dragon curve with rounded corners

Chapter 4

UNWIND Evolute circle

ARCHI Archimedes spiral

LOGSPIRA Logarithmic spiral

SPHERSPI Spherical spirals

WHIRL Rotating and shrinking square

PYTHT1 Pythagoras tree, number system method

PYTHT2 Lopsided Pythagoras tree

PYTHT3 Pythagoras tree, backtrack method

PYTHB Bare branching Pythagoras tree

TREEM Mandelbrot tree, backtrack method

STAR Star fractal

Chapter 5

DUSTB Dust fractal of two rotations, backtrack method

DUSTBT Dust fractal of three rotations, backtrack method

Chapter 6

MONDRIAN Pattern of random horizontal and vertical lines

DUST Dust fractal of two rotations or mirrorings, Monte Carlo method

PYTHTD Pythagoras tree with random disturbances, backtrack method

BROWNL Brownian line

COLLET Bifurcation diagram of $x \rightarrow ax(1-x)$

Chapter 7

HENON Orbits of Hénon's quadratic system

MIRA Orbits of Mira's dynamical system

JULIAB Julia fractal of quadratic system, backtrack method

MANDEL Mandelbrot set of quadratic system, pixel method

Chapter 8

CLOUD Orbits of dynamical system

CLOUD1 Orbits of dynamical system

CLOUD2 Orbits of dynamical system

DUSTV Dust fractal, Monte Carlo method

MEANDERN Meandering line with given starting line and motif, number system method

MEANDERB Meandering line with given starting line and motif, backtrack method

ARCHI

```
10 REM ***ARCHIMEDES SPIRAL***
20 REM ***NAME:ARCHI***
30 SCREEN 3 : CLS : PI=3.141593
40 WINDOW (-12,-9)-(12,9)
50 A=.1 : PSET (0,0)
60 FOR T=0 TO 16*PI STEP .1  : R=A*T
70 X=R*COS(T) : Y=R*SIN(T)
80 LINE -(X,Y)
90 NEXT T
100 A$=INPUT$(1) : END
```

BROWNL

```
10 REM ***BROWNIAN LINE***
20 REM ***NAME:BROWNL***
30 SCREEN 3 : CLS : RANDOMIZE TIMER
40 WINDOW (-1.2,-.9)-(1.2,.9)
50 W=40 : Y=0
60 LINE (1,0)-(-1,0)
70 FOR K=0 TO 2000
80 X=-1+K/1000 : Y=Y+W*(RND-.5)/2000
90 LINE -(X,Y)
100 NEXT K : A$=INPUT$(1) : END
```

CLOUD

```
10 REM ***ORBITS OF A DYNAMICAL SYSTEM***
20 REM ***NAME:CLOUD***
30 SCREEN 3 : CLS
40 WINDOW (-120,-90)-(120,90)
50 A=3.5 : B=-3
60 X=3.21 : Y=6.54 : GOSUB 110
70 FOR N=0 TO 10000 : PSET (X,Y)
80 Z=X : X=Y+W
90 GOSUB 110
100 Y=W-Z : NEXT N : BEEP : A$=INPUT$(1) : END
110 IF X>1 THEN W=A*X+B*(X-1) : RETURN
120 IF X<-1 THEN W=A*X+B*(X+1) : RETURN
130 W=A*X : RETURN : END
140 REM ***ALTERNATIVES FOR GOSUB***
150 REM ***TRY VARIOUS A,B VALUES EACH WITH A NUMBER OF
STARTING VALUES X,Y***
160 W=A*X+B*SIN(X) : RETURN
170 W=A*X+B*COS(X) : RETURN
180 W=A+B*SIN(X) : RETURN
190 W=A+B*COS(X) : RETURN
200 IF ABS(X)<1 THEN W=A*X ELSE W=B*X+(A-B)/X
210 RETURN : END
```

CLOUD1

```
10 REM ***ORBITS OF A DYNAMICAL SYSTEM***
20 REM ***NAME:CLOUD1***
30 SCREEN 3 : CLS
40 DIM X(8),Y(8),P(8)
50 WINDOW (-40,-30)-(40,30)
60 A=-.5 : B=2 : REM ***PARAMETER VALUES***
70 DATA 2,0,200,4,0,400,6,0,600,8,0,800
80 DATA 10,0,1000,12,0,1200,14,0,1400,16,0,1600
90 FOR I=1 TO 8 : READ X(I),Y(I),P(I) : NEXT I
100 FOR K=1 TO 8 : X=X(K) : Y=Y(K) : P=P(K)
110 GOSUB 140 : FOR N=0 TO P : PSET (X,Y)
120 Z=X : X=Y+W : GOSUB 140 : Y=W-Z
130 NEXT N : NEXT K  : BEEP : A$=INPUT$(1) : END
140 W=X*(A+B/(1+ABS(X))) : RETURN : END
```

CLOUD2

```
10 REM ***ORBITS OF A DYNAMICAL SYSTEM***
20 REM ***NAME:CLOUD2***
30 SCREEN 3 : CLS : PI=3.141593
40 WINDOW (-8 ,-5 )-(8 ,7)
50 REM ***PARAMETER VALUES***
60 A=.51 : B=-.49 : C=.9995
70 X=4  : Y=1  : GOSUB 110
80 FOR N=0 TO 9000 : PSET (Y,X)
90 Z=X : X=C*Y+W : GOSUB 110 : Y=W-Z
100 NEXT N : BEEP : A$=INPUT$(1) : END
110 IF X>0 THEN W=A*X ELSE W=B*X
120 RETURN : END
```

COLLET

```
10 REM ***BIFURCATION OF THE VERHULST MODEL***
20 REM ***WITH THE BIFURCATION DIAGRAM AFTER COLLET***
30 REM ***NAME:COLLET***
40 SCREEN 1 : CLS : LOCATE 10,1
50 PRINT "ITERATION OF X:=AX(1-X) FOR SUCCESSIVE"
60 LOCATE 12,14 : PRINT "VALUES OF A"
70 FOR I=1 TO 10000 : NEXT I
80 REM ***ITERATION OF A*X*(1-X) FOR SUCCESSIVE VALUES
    OF A***
90 SCREEN 3 : CLS
100 WINDOW (-.1,-.1)-(1.5,1.1)
110 FOR I=1 TO 12 : CLS
120 A=2.8+I/10 : X=.01 : M=40 : X0=X
130 PSET (0,0) : FOR K=0 TO 100
140 LINE -(K/100,A*K*(1-K/100)/100)
150 NEXT K
160 LINE (0,0)-(1,0) : LINE (0,0)-(1,1)
170 LOCATE 4,2
180 PRINT "A =";A
```

```
190 PSET(X,0) : FOR N=0 TO M+5*I
200 Y=A*X*(1-X)
210 LINE -(X,Y) : LINE -(Y,Y)
220 X=Y : NEXT N
230 FOR J=1 TO 10000
240 NEXT J : NEXT I
250 SCREEN 1 : CLS : LOCATE 8,4
260 PRINT "BIFURCATION DIAGRAM OF X:=AX(1-X)"
270 LOCATE 12,10
280 PRINT "A RUNS FROM 2.8 TO 4"
290 FOR I=1 TO 10000 : NEXT I
300 REM ***BIFURCATION DIAGRAM  OF X:=AX(1-X) ***
310 SCREEN 3 : CLS
320 LINE (40,390)-(600,390)
330 FOR I=0 TO 12
340 LINE (40+I*140/3,390)-(40+I*140/3,385)
350 NEXT I
360 FOR N=0 TO 280 STEP 2
370 A=2.8+1.2*N/280 : X=.7
380 FOR K=1 TO 150
390 X=A*X*(1-X)
400 IF K>50 THEN  PSET(40+2*N,300-250*X)
410 NEXT K : NEXT N
420 FOR I=1 TO 10000 : NEXT I
430 REM ***PERIOD DOUBLING WINDOW"
440 SCREEN 1 : CLS : LOCATE 10,9
450 PRINT "PERIOD DOUBLING WINDOW"
460 LOCATE 12,12
470 PRINT " 3.44 < A < 3.6 "
480 FOR I=1 TO 10000 : NEXT I
490 SCREEN 3 : CLS
500 FOR N=0 TO 280 STEP 1.5
510 A=3.4+.2*N/280 : X=.7
520 FOR K=1 TO 160
530 X=A*X*(1-X)
540 IF K>80 THEN  PSET(40+2*N,540-600*X)
550 NEXT K : NEXT N
560 A$=INPUT$(1) : END
```


```
COMB

10 REM ***CANTOR COMB***
20 REM ***NAME:COMB***
30 SCREEN 3 : CLS
40 WINDOW (-.3,-.4)-(1.3,.8)
50 DIM A(729),B(729) : A(0)=0 : A(1)=1
60 B=.1 : LINE (0,0)-(1,0) : LINE -(1,-B) : LINE -(0,-B) :
   LINE -(O,0)
70 FOR P=1 TO 6
80 FOR I=0 TO 2^P-1
90 B(I)=A(I)/3 : B(I+2^P)=1-(1-A(I))/3
100 NEXT I
110 FOR J=1 TO 2^(P+1)-1
120 A(J)=B(J) : NEXT J
130 FOR K=0 TO 2^(P+1)-1 STEP 2
140 LINE (A(K),B*P)-(A(K+1),B*P)
```

```
150 LINE (A(K),B*P)-(A(K),B*P-B)
160 LINE (A(K+1),B*P)-(A(K+1),B*P-B)
170 NEXT K : NEXT P
180 BEEP : A$=INPUT$(1) : END
```

DRAGON

```
10 REM ***DRAGON CURVE WITH ARBITRARY ANGLE***
20 REM ***NAME:DRAGON***
30 SCREEN 3 : CLS : PI=3.141593
40 WINDOW (-4,-3)-(4,3) : REM ***ADAPT IF NECESSARY***
50 PRINT "SELECT ORDER P AS AN INTEGER LESS THAN 11***
60 INPUT P : CLS : H=2^(-P/2)
70 PRINT "SELECT ANGLE IN DEGREES, TAKE A=90 OR SLIGHTLY
   LARGER"
80 INPUT A : CLS : A=A*PI/180
90 B=PI-A : X=H : Y=0 : LINE (0,0)-(H,0) : S=0
100 FOR N=1 TO 2^P-1 : M=N
110 IF M MOD 2 = 0 THEN M=M/2 : GOTO 110
120 IF M MOD 4 = 1 THEN D=1 ELSE D=-1
130 S=(S+D)
140 X=X+H*COS(S*B)
150 Y=Y+H*SIN(S*B)
160 LINE -(X,Y) : NEXT N
170 BEEP : A$=INPUT$(1) : END
```

DRAGON0

```
10 REM ***DRAGON CURVE BETWEEN TWO POINTS***
20 REM ***NAME:DRAGON0***
30 SCREEN 3 : CLS : PI=3.141593
40 WINDOW (-.7,-1)-(1.7,.8)
50 P=12 : REM ***CHOICE OF ORDER***
60 H=2^(-P/2) : S=0
70 X=H*COS(P*PI/4) : Y=H*SIN(P*PI/4)
80 LINE (0,0)-(X,Y)
90 FOR N=1 TO 2^P-1 : M=N
100 IF M MOD 2 = 0 THEN M=M/2 : GOTO 100
110 IF M MOD 4 = 1 THEN D=1 ELSE D=-1
120 S=(S+D) : F=(S-P/2)*PI/2
130 X=X+H*COS(F) : Y=Y+H*SIN(F)
140 LINE -(X,Y)
150 NEXT N : A$=INPUT$(1) : END
```

DRAGON1

```
10 REM ***DRAGON CURVE WITH ROUNDED CORNERS***
20 REM ***NAME:DRAGON1***
30 SCREEN 3 : CLS : PI=3.141593
40 WINDOW (-.7,-1.1)-(1.7,.7)
```

```
50 P=10 : REM ***ORDER***
60 H=2^(-P/2) : S=0
70 X1=H*COS(P*PI/4) : Y1=-H*SIN(P*PI/4)
80 LINE (0,0)-(.75*X1,.75*Y1)
90 FOR N=1 TO 2^P-1 : M=N
100 IF M MOD 2 = 0 THEN M=M/2 : GOTO 100
110 IF M MOD 4 = 1 THEN D=1 ELSE D=-1
120 S=(S+D) MOD 4
130 X2=X1+H*COS((S-P/2)*PI/2)
140 Y2=Y1+H*SIN((S-P/2)*PI/2)
150 XA=(3*X1+X2)/4 : YA=(3*Y1+Y2)/4
160 XB=(X1+3*X2)/4 : YB=(Y1+3*Y2)/4
170 LINE -(XA,YA) : LINE -(XB,YB)
180 X1=X2 : Y1=Y2 : NEXT N
190 LINE -(1, 0) : BEEP
200 A$=INPUT$(1) : END
```

DUST

```
10 REM ***DUST FRACTAL, MONTE CARLO METHOD***
20 REM ***NAME:DUST***
30 SCREEN 3 : CLS : PI=3.141593 : RANDOMIZE 100
40 WINDOW (-1.1,-1.2)-(2.1,1.2)
50 R1=.6 : A=R1*COS(2*PI/3) : B=R1*SIN(2*PI/3)
60 R2=.6 : C=R2*COS(2*PI/3) : D=-R2*SIN(2*PI/3)
70 X=A : Y=B : REM ***COORDINATES STARTING POINT***
80 FOR K=1 TO  10000
90 IF RND <.5 THEN GOSUB 120 ELSE GOSUB 130
100 PSET (X,Y) : NEXT K
110 BEEP : A$=INPUT$(1) : END
120 Z=X : X=A*X-B*Y : Y=B*Z+A*Y : RETURN
130 Z=X : X=C*X-D*Y+1-C : Y=D*Z+C*Y-D : RETURN
140 END
```

DUSTB

```
10 REM ***DUST FRACTAL, BACKTRACKING***
20 REM ***NAME:DUSTB***
30 SCREEN 3 : CLS
40 WINDOW (-1.1,-1)-(2.1,1.4)
50 P=11 : DIM X1(P),Y1(P),X2(P),Y2(P)
60 A=.5 : B=.5 : C=.5 : D=-.5 : REM ***TRANSFORMATION
   PARAMETERS***
70 X1(0)=A : Y1(0)=B
80 PSET (0,0) : PSET (1,0) : PSET (A,B)
90 S=1 : GOSUB 140
100 FOR M=1 TO 2^(P-1)-1 : S=P : N=M
110 IF N MOD 2=0  THEN N=N\2 : S=S-1 : GOTO 110
120 GOSUB 140 : NEXT M : BEEP
130 A$=INPUT$(1) : END
140 X1(S-1)=X2(S-1) : Y1(S-1)=Y2(S-1)
150 FOR J=S TO P
160 X=X1(J-1) : Y=Y1(J-1)
```

```
170 X1(J)=A*X-B*Y : Y1(J)=B*X+A*Y
180 X2(J)=C*X-D*Y+1-C : Y2(J)=D*X+C*Y-D
190 PSET (X1(J),Y1(J)) : PSET (X2(J),Y2(J))
200 NEXT J : RETURN : END
```

DUSTBT

```
10 REM ***DUST FRACTAL OF THREE ROTATIONS, BACKTRACKING***
20 REM ***NAME:DUSTBT***
30 SCREEN 3 : CLS
40 WINDOW (-.8 ,-.6 )-(1.6,1.2)
50 P=7 : DIM X1(P),Y1(P),X2(P),Y2(P),X3(P),Y3(P)
60 T1=.5 : T2=.866 : REM ***POSITION TOP***
70 A=.43 : B=.3 : C=A : D=B : E=A : F=B
80 G=T1*(1-E)+T2*F : H=-T1*F+T2*(1-E)
90 PSET (0,0) : PSET (1,0) : PSET (T1,T2)
100 X1(0)=.5 : Y1(0)=.289 : PSET (X1(0),Y1(0))
110 FOR M=0 TO 3^(P-1)-1 : S=P : N=M
120 IF M=0 THEN S=1 : GOTO 160
130 IF N MOD 3=0 THEN N=N\3 : S=S-1 : GOTO 130
140 X1(S-1)=X2(S-1) : Y1(S-1)=Y2(S-1)
150 X2(S-1)=X3(S-1) : Y2(S-1)=Y3(S-1)
160 FOR J=S TO P
170 X=X1(J-1) : Y=Y1(J-1)
180 X1(J)=A*X-B*Y : Y1(J)=B*X+A*Y
190 X2(J)=C*X-D*Y+1-C : Y2(J)=D*X+C*Y-D
200 X3(J)=E*X-F*Y+G : Y3(J)=F*X+E*Y+H
210 PSET (X1(J),Y1(J)) : PSET (X2(J),Y2(J)) :
    PSET (X3(J),Y3(J))
220 NEXT J : NEXT M : BEEP : A$=INPUT$(1) : END
```

DUSTV

```
10 REM ***DUST FRACTAL, VARIA, MONTE CARLO METHOD***
20 REM ***NAME:DUSTV***
30 SCREEN 3 : CLS : PI=3.141593 : RANDOMIZE 100
40 WINDOW (-3.1,-2.7)-(4.1,2.7)
50 R=.7 : A=R*COS(2*PI/3) : B=R*SIN(2*PI/3) : C=2.5 : D=.9
60 X=1 : Y=0 : REM ***COORDINATES STARTING POINT***
70 FOR K=1 TO 10000
80 IF RND <.5 THEN GOSUB 110 ELSE GOSUB 130
90 PSET (X,Y) : NEXT K
100 BEEP : A$=INPUT$(1) : END
110 Z=X : X=A*X-B*Y : Y=B*Z+A*Y : RETURN
120 Z=X : X=D*X+Y+1-D : Y=-Z-D*Y+1 : RETURN
130 Z=X : X=.2*(X-1)*(X-1)-Y+1 : Y=.8*Z : RETURN
140 Z=X : X=(X*X-Y*Y+C-1)/C : RETURN
150 Z=X : X=(X*X+Y*Y+C-1)/C : Y=2*Z*Y/C : RETURN
160 END
```

HENON

```
10 REM ***ORBITS OF HENON'S QUADRATIC SYSTEM***
20 REM ***NAME:HENON***
30 SCREEN 3 : CLS : RANDOMIZE 1
40 FOR K=1 TO 20
50 X=-.4+RND : Y=-.4+RND
60 WINDOW (-1.6,-1.2)-(1.6,1.2)
70 A=.24 : B=SQR(1-A^2)
80 FOR N=1 TO 500  : PSET (X,Y)
90 Z=X : X=X*A-(Y-X*X)*B
100 Y=Z*B+(Y-Z*Z)*A
110 IF ABS(X)+ABS(Y)>10 THEN GOTO 130
120 NEXT N
130 NEXT K : A$=INPUT$(1) : END
```

JULIAB

```
10 REM ***JULIA FRACTAL OF Z:=Z^2+C , BACKTRACKING***
20 REM ***NAME:JULIAB***
30 SCREEN 3 : CLS
40 WINDOW (-2,-1.5)-(2,1.5)
50 P=12 : DIM X1(P),Y1(P),X2(P),Y2(P)
60 A=.25 : B=.25
70 A1=(.25-A)/2 : B1=-B/2 : R1=SQR(A1^2+B1^2)
80 X1(0)=.5+SQR(R1+A1) : Y1(0)=SQR(R1-A1)
90 IF B>0 THEN Y1(0)=-Y1(0)
100 PSET (X1(0),Y1(0)) : S=1 : GOSUB 150
110 FOR M=1 TO 2^(P-1)-1 : S=P : N=M
120 IF N MOD 2 =0 THEN N=N\2 : S=S-1 : GOTO 120
130 GOSUB 140 : NEXT M : BEEP : END
140 X1(S-1)=X2(S-1) : Y1(S-1)=Y2(S-1)
150 FOR J=S TO P
160 X=X1(J-1) : Y=Y1(J-1)
170 R=SQR((X-A)^2+(Y-B)^2)/2 : T=(X-A)/2
180 X1(J)=SQR(R+T) : X2(J)=-X1(J)
190 Y1(J)=SQR(R-T) : IF Y<B THEN Y1(J)=-Y1(J)
200 Y2(J)=-Y1(J)
210 PSET (X1(J),Y1(J)) : PSET (X2(J),Y2(J))
220 NEXT J : RETURN : END
```

KOCH

```
10 REM ***KOCH CURVE***
20 REM ***NAME:KOCH***
30 SCREEN 3 : CLS : PI=3.141593
40 WINDOW (-.1,-.4)-(1.1,.5)
50 P=4 : DIM T(P) : REM ***ORDER***
60 H=3^(-P) : PSET (0,0)
70 FOR N=0 TO 4^P-1
80 REM ***QUATERNARY NOTATION OF N***
90 M=N : FOR L=0 TO P-1
100 T(L)=M MOD 4 : M=M\4 : NEXT L
```

```
110 REM ***DETERMINATION SLOPE OF NTH LINE SEGMENT***
120 S=0 : FOR K=0 TO P-1
130 S=S+(T(K)+1) MOD 3-1
140 NEXT K
150 REM ***GRAPH OF NTH LINE SEGMENT***
160 X=X+COS(PI*S/3)*H
170 Y=Y+SIN(PI*S/3)*H
180 LINE -(X,Y)
190 NEXT N : BEEP : A$=INPUT$(1) : END
```

LEVY

```
10 REM ***PYTHAGORAS OR LEVY CURVE***
20 REM ***NAME:LEVY***
30 SCREEN 3 : CLS : PI=3.141593
40 WINDOW (-.7, -1.3)-(1.7,.5)
50 PRINT "SELECT ORDER P AS A SMALL INTEGER LESS THAN 12"
60 INPUT P : CLS
70 H=2^(-(P/2)) : A=H*COS(P*PI/4 ): B=H*SIN(P*PI/4)
80 LINE (0,0)-(A,-B) : LINE -(A+B,A-B)
90 X=1 :Y=1
100 FOR N=2 TO 2^P-1
110 M=N : S=1
120 IF M MOD 2=1 THEN S=S+1
130 M=M \2
140 IF M>1 THEN GOTO 120
150 IF S MOD 4=0 THEN X=X+1
160 IF S MOD 4=1 THEN Y=Y+1
170 IF S MOD 4=2 THEN X=X-1
180 IF S MOD 4=3 THEN Y=Y-1
190 LINE -(A*X+B*Y,A*Y-B*X)
200 NEXT N : BEEP : A$=INPUT$(1) : END
```

LOGSPIRA

```
10 REM ***LOGARITHMIC SPIRAL***
20 REM ***NAME:LOGSPIRA***
30 SCREEN 3 : CLS : PI=3.141593
40 WINDOW (-4,-3)-(4,3)
50 A=.05 : B=.1 : REM ***STARTING POINT AND GROWTH RATE***
60 PSET (A,0)
70 FOR T=0 TO 42 STEP .1 : R=A*EXP(B*T)
80 X=R*COS(T) : Y=R*SIN(T)
90 LINE -(X,Y)
100 NEXT T : A$=INPUT$(1) : END
```

MANDEL

```
10 REM ***MANDELBROT SET, TOTAL VIEW***
20 REM ***NAME:MANDEL***
```

```
30 SCREEN 3 : CLS
40 WINDOW (-2.2,-1.4)-(1.1,1.4)
50 N1=160 : N2=100 : REM ***RESOLUTION***
60 FOR I=-N1 TO N1 : A=-.55+1.65*I/N1
70 FOR J=0 TO N2 : B=1.4*J/N2
80 U=4*(A*A+B*B) : V=U-2*A+1/4
90 IF U+8*A+15/4<0 THEN K=1 : GOTO 170
100 IF V-SQR(V)+2*A-1/2<0 THEN K=1 : GOTO 170
110 X=A : Y=B
120 FOR K=1 TO 50
130 U=X*X : V=Y*Y : W=2*X*Y
140 X=U-V+A : Y=W+B
150 IF U+V>16 THEN GOTO 170
160 NEXT K
170 L=K MOD 2 : PSET (A,B),L : PSET (A,-B),L
180 NEXT J : NEXT I
190 BEEP : A$=INPUT$(1)
```

MEANDER

```
10 REM ***FRACTAL CURVE WITH A GIVEN BASE LINE AND MOTIF,
   KOCH CROSS***
20 REM ***NAME:MEANDER***
30 SCREEN 3 : CLS : PI=3.141593
40 DIM X(2048),Y(2048)
50 WINDOW (-2.4,-1.8)-(2.4,1.8)
60 U=4 : DIM A(U),B(U) : REM ***NUMBER ELEMENTS BASE LINE***
70 V=4 : DIM C(V),D(V) : REM ***NUMBER ELEMENTS MOTIF***
80 DATA 1,1,-1,1,-1,-1,1,-1,1,1 : REM ***DATA BASE LINE***
90 DATA .333,0,.5,.2887,.667,0 : REM ***DATA MOTIF***
100 PRINT "SELECT ORDER P AS A SMALL INTEGER"
110 INPUT P : CLS
120 FOR I=0 TO U : READ A(I),B(I) : NEXT I
130 FOR I=1 TO V-1 : READ C(I),D(I) : NEXT I
140 REM ***DETERMINATION COORDINATES VERTICES OF MEANDERING
    LINE***
150 C(0)=0 : D(0)=0 : X(0)=0 : Y(0)=0 : X(V^P)=1 : Y(V^P)=0
160 FOR I=0 TO P-1
170 FOR J=0 TO V^P-1 STEP V^(P-I)
180 M1=J+V^(P-I) : X1=X(M1)-X(J) : Y1=Y(M1)-Y(J)
190 FOR K=1 TO V-1
200 M2=J+K*V^(P-I-1)
210 X(M2)=X1*C(K)-Y1*D(K)+X(J) : Y(M2)=Y1*C(K)+X1*D(K)+Y(J)
220 NEXT K : NEXT J : NEXT I
230 REM ***GRAPHICS, FOR EACH SIDE OF THE BASE LINE THE
    MEANDER IS DRAWN***
240 PSET (A(0),B(0))
250 FOR M=0 TO U-1 : A=A(M+1)-A(M)
260 B=B(M+1)-B(M) : FOR N=0 TO V^P
270 X=A*X(N)-B*Y(N)+A(M) : Y=B*X(N)+A*Y(N)+B(M)
280 LINE - (X,Y)
290 NEXT N : NEXT M
300 A$=INPUT$(1) : END
```

MEANDERB

```
10 REM ***FRACTAL LINE WITH A GIVEN MOTIF, BACKTRACKING***
20 REM ***NAME:MEANDERB***
30 SCREEN 3 : CLS : PI=3.141593
40 WINDOW (-3.2,-2.4)-(3.2,2.4)
50 U=4 : DIM A(U),B(U) : REM ***NUMBER ELEMENTS OF BASE
   LINE***
60 REM ***BASE LINE, VECTORS IN TRUE LENGTH AND DIRECTION IN
   DEGREES***
70 DATA 2,0,2,90,2,180,2,-90
80 X0=-1 : Y0=-1 : REM ***COORDINATES STARTING POINT***
90 XM=X0 : YM=Y0
100 FOR I=0 TO U-1 : READ A(I),B(I)
110 B(I)= B(I)*PI/180 : NEXT I
120 V=4 : DIM C(V),D(V),L(V),F(V) : REM ***NUMBER ELEMENTS OF
    MOTIF***
130 REM ***MOTIF, VECTORS IN RELATIVE LENGTH AND DIRECTION IN
    DEGREES***
140 DATA 1,0,1,60,1,-60,1,0
150 FOR I=1 TO V : READ L(I),F(I)
160 F(I)=F(I)*PI/180 : NEXT I
170 S=0 : FOR I=1 TO V : S=S+L(I)*COS(F(I)) : NEXT I
180 FOR I=1 TO V : L(I)=L(I)/S
190 C(I)=L(I)*COS(F(I)) : D(I)=L(I)*SIN(F(I)) : NEXT I
200 PRINT "TAKE ORDER P AS A SMALL INTEGER"
210 INPUT P : DIM X(V,P),Y(V,P) : CLS
220 PSET (X0,Y0)
230 FOR Q=0 TO U-1 : X(1,0)=1 : Y(1,0)=0 : S=1
240 A=A(Q)*COS(B(Q)) : B=A(Q)*SIN(B(Q)) : GOSUB 310
250 FOR M=1 TO V^(P-1)-1 : N=M : S=P
260 IF N MOD V=0 THEN N=N\V : S=S-1 : GOTO 260
270 GOSUB 290 : NEXT M : NEXT Q
280 BEEP : A$=INPUT$(1) : END
290 FOR I=1 TO V-1
300 X(I,S-1)=X(I+1,S-1) : Y(I,S-1)=Y(I+1,S-1) : NEXT I
310 FOR J=S TO P : FOR K=1 TO V
320 X(K,J)=C(K)*X(1,J-1)-D(K)*Y(1,J-1)
330 Y(K,J)=D(K)*X(1,J-1)+C(K)*Y(1,J-1)
340 NEXT K : NEXT J
350 FOR T=1 TO V
360 XS=A*X(T,P)-B*Y(T,P) : YS=B*X(T,P)+A*Y(T,P)
370 XM=XM+XS : YM=YM+YS : LINE -(XM,YM)
380 NEXT T : RETURN : END
```

MEANDERN

```
10 REM ***FRACTAL CURVE, NUMBER SYSTEM METHOD, KOCH CROSS***
20 REM ***NAME:MEANDERN***
30 SCREEN 3 : CLS : PI=3.141593
40 WINDOW (-2.8,-2.1)-(2.8,2.1)
50 P=4 : DIM T(P) : REM ***ORDER***
60 U=4 : DIM A(U),B(U) : REM ***NUMBER ELEMENTS OF BASE
   LINE***
70 REM ***BASE LINE, VECTORS IN LENGTH AND DIRECTION IN
   DEGREES***
```

```
80 DATA 2,0,2,90,2,180,2,-90
90 X0=-1 : Y0=-1 : REM ***COORDINATES STARTING VERTEX***
100 XM=X0 : YM=Y0
110 FOR I=0 TO U-1 : READ A(I),B(I)
120 B(I)=B(I)*PI/180 : NEXT I
130 V=4 : DIM L(V),F(V) : REM ***NUMBER ELEMENTS OF MOTIF***
140 REM ***MOTIF, VECTORS IN LENGTH AND DIRECTION IN
    DEGREES***
150 DATA 1,0,1,60,1,-60,1,0
160 FOR I=0 TO V-1 : READ L(I),F(I)
170 F(I)=F(I)*PI/180 : NEXT I
180 S=0 : FOR I=0 TO V-1 : S=S+L(I)*COS(F(I)) : NEXT I
190 FOR I=0 TO V-1 : L(I)=L(I)/S : NEXT I
200 PSET (X0,Y0) : FOR K=0 TO U-1
210 FOR N=0 TO V^P-1 : REM ***MAIN LOOP***
220 M=N : FOR J=0 TO P-1
230 T(J)=M MOD V : M=M\V : NEXT J
240 L=A(K) : F=B(K)
250 FOR J=0 TO P-1
260 L=L*L(T(J)) : F=F+F(T(J)) : NEXT J
270 XS=L*COS(F) : YS=L*SIN(F)
280 XM=XM+XS : YM=YM+YS : LINE -(XM,YM)
290 NEXT N : NEXT K
300 A$=INPUT$(1) : END




MINK

10 REM ***MINKOWSKI SAUSAGE***
20 REM ***NAME:MINK***
30 SCREEN 3 : CLS
40 WINDOW (-.3,-.7)-(1.3,.5)
50 DIM A(7) : A(0)=0 : A(1)=1 : A(2)=0 : A(3)=3
60 A(4)=3 : A(5)=0 : A(6)=1 : A(7)=0
70 P=3 : DIM T(P) : REM ***ORDER***
80 H=4^(-P) : X=0 : Y=0 : PSET (0,0)
90 FOR N=0 TO 8^P-1
100 M=N : FOR L=0 TO P-1
110 T(L)=M MOD 8 : M=INT(M/8) : NEXT L
120 S=0 : FOR K=0 TO P-1
130 S=S+A(T(K)) : S=S MOD 4
140 NEXT K
150 IF S=0 THEN X=X+H
160 IF S=1 THEN Y=Y+H
170 IF S=2 THEN X=X-H
180 IF S=3 THEN Y=Y-H
190 LINE -(X,Y)
200 NEXT N : BEEP
210 A$=INPUT$(1) : END




MIRA

10 REM ***ORBITS OF MIRA'S SYSTEM***
20 REM ***NAME:MIRA***
```

```
30 SCREEN 3 : CLS
40 WINDOW (-20,-15)-(20,15)
50 A=.7 : B=.9998 : P=12000 : C=2-2*A
60 X=0 : Y=12.1 : REM ***STARTING POINT***
70 W=A*X+C*X*X/(1+X*X)
80 FOR N=0 TO P
90 IF N>100 THEN PSET (X,Y)
100 Z=X : X=B*Y+W : U=X*X
110 W=A*X+C*U/(1+U) : Y=W-Z
120 NEXT N
130 A$=INPUT$(1) : END
```

MONDRIAAN

```
10 REM ***MODERN ART***
20 REM ***NAME:MONDRIAAN***
30 SCREEN 3 : CLS : PI=3.141593 : RANDOMIZE 111
40 WINDOW (-.5,-.3)-(1.5,1.2)
50 DIM X(200),Y(200) : FOR I=1 TO 200 : X(I)=RND : Y(I)=RND :
   NEXT I
60 PRINT "THIS PROGRAM FORMS WITH A GIVEN MOTIF OF RANDOM
POSITION"
70 PRINT "AND SIZE A MODERN ART COMPOSITION"
80 PRINT "THE FOLLOWING ORNAMENTS ARE AVAILABLE"
90 PRINT "KEY 1 FOR  HORIZONTAL AND VERTICAL LINES"
100 PRINT "KEY 2 FOR EQUILATERAL TRIANGLES IN FIXED
    ORIENTATION"
110 PRINT "KEY 3 FOR EQUILATERAL TRIANGLES IN RANDOM
    ORIENTATION"
120 PRINT "KEY 4 FOR SQUARES"
130 PRINT "KEY 5 FOR PENTAGRAMS"
140 PRINT "KEY 6 FOR CIRCLES"
150 PRINT "KEY 7 VOOR END"
160 INPUT S : CLS
170 IF S=7 THEN CLS : END
180 PRINT "SELECT NUMBER N OF ORNAMENTS AND SELECT SCALE
FACTOR H"
190 PRINT "TAKE FOR N AN INTEGER <=200 AND FOR H AN INTEGER
    BETWEEN 1 AND 9"
200 PRINT "H=9 GIVES THE LARGEST SCALE"
210 INPUT N,H : CLS
220 WINDOW (-.5,-.3)-(1.5,1.2)
230 FOR K=1 TO N
240 V=H*(1-SQR(RND))/50
250 ON S GOSUB 270,290,320,360,380,450
260 NEXT K : A$=INPUT$(1) : CLS : GOTO 60
270 IF RND >.5 THEN LINE (X(K)-V,Y(K))-(X(K)+V,Y(K)) ELSE
LINE (X(K),Y(K)-V)-(X(K),Y(K)+V)
280 RETURN
290 B=2*PI/3 : U1=X(K)+V : V1=Y(K)
300 U2=X(K)+V*COS(B) : V2=Y(K)+V*SIN(B) :
    U3=U2 : V3=Y(K)-V*SIN(B)
310 LINE (U1,V1)-(U2,V2) : LINE -(U3,V3) : LINE -(U1,V1) :
RETURN
320 B=2*PI/3 : A=B*RND : U1=X(K)+V*COS(A) : V1=Y(K)+V*SIN(A)
330 U2=X(K)+V*COS(A+B) : V2=Y(K)+V*SIN(A+B)
```

```
340 U3=X(K)+V*COS(A-B) : V3=Y(K)+V*SIN(A-B)
350 LINE (U1,V1)-(U2,V2) : LINE -(U3,V3) : LINE -(U1,V1) :
    RETURN
360 W=V/2 : LINE (X(K)+W,Y(K)+W)-(X(K)-W,Y(K)+W) :
    LINE -(X(K)-W,Y(K)-W)
370 LINE -(X(K)+W,Y(K)-W) : LINE -(X(K)+W,Y(K)+W) : RETURN
380 B=2*PI/5 : U1=X(K)+V : V1=Y(K)
390 U2=X(K)+V*COS(B) : V2=Y(K)+V*SIN(B)
400 U3=X(K)+V*COS(2*B) : V3=Y(K)+V*SIN(2*B)
410 U4=X(K)+V*COS(3*B) : V4=Y(K)+V*SIN(3*B)
420 U5=X(K)+V*COS(4*B) : V5=Y(K)+V*SIN(4*B)
430 LINE (U1,V1)-(U3,V3) : LINE -(U5,V5) : LINE -(U2,V2)
440 LINE -(U4,V4) : LINE -(U1,V1) : RETURN
450 CIRCLE (X(K),Y(K)),V : RETURN


PYTHB

10 REM ***BARE PYTHAGORAS TREE***
20 REM ***NAME:PYTHB***
30 SCREEN 3 : CLS : PI=3.141593
40 WINDOW (-3.5,-2)-(4.5,4)
50 DIM X(2048),Y(2048)
60 REM ***CHOICE OF ANGLE***
70 F=PI/4 : C=COS(F) : S=SIN(F)
80 A1=-C*S : A2=C^2 : B1=A1+A2 : B2=-A1+A2
90 C1=B2 : C2=1-B1 : D1=1-A1 : D2=1-A2
100 X(2)=0 : Y(2)=0 : X(3)=1 : Y(3)=0
110 LINE (.5,-1)-(.5,0)
120 FOR M=1 TO 9
130 FOR J=0 TO 2^(M-1)-1
140 X0=X(2^M+2*J) : Y0=Y(2^M+2*J)
150 X1=X(2^M+2*J+1) : Y1=Y(2^M+2*J+1)
160 U=X1-X0 : V=Y1-Y0
170 XA=X0+A1*U-A2*V : YA=Y0+A2*U+A1*V
180 XB=X0+B1*U-B2*V : YB=Y0+B2*U+B1*V
190 XC=X0+C1*U-C2*V : YC=Y0+C2*U+C1*V
200 XD=X0+D1*U-D2*V : YD=Y0+D2*U+D1*V
210 X(2^(M+1)+4*J)=XA : Y(2^(M+1)+4*J)=YA
220 X(2^(M+1)+4*J+1)=XB : Y(2^(M+1)+4*J+1)=YB
230 X(2^(M+1)+4*J+2)=XC : Y(2^(M+1)+4*J+2)=YC
240 X(2^(M+1)+4*J+3)=XD : Y(2^(M+1)+4*J+3)=YD
250 LINE ((X0+X1)/2,(Y0+Y1)/2)-((XA+XB)/2,(YA+YB)/2)
260 LINE ((X0+X1)/2,(Y0+Y1)/2)-((XC+XD)/2,(YC+YD)/2)
270 NEXT J : NEXT M : BEEP
280 A$=INPUT$(1) : END


PYTHT1

10 REM ***PYTHAGORAS TREE***
20 REM ***USING BINARY NUMBER SYSTEM***
30 REM ***NAME:PYTHT1***
40 SCREEN 3 : CLS : PI=3.141593
50 WINDOW (-8,-4)-(8,8)
```

```
60 P=8 : DIM A(P) : REM ***ORDER***
70 X=0 : Y=0 : U=1 : V=1 : C=1/SQR(2)
80 FOR M=0 TO P
90 FOR N=2^M TO 2^(M+1)-1
100 L=N : H=1 : X=0 : Y=0 : F=0
110 FOR K=0 TO M-1
120 A(M-K)=L MOD 2 : L=INT(L/2) : NEXT K
130 X=0 : Y=0
140 FOR J=1 TO M
150 IF A(J)=0 THEN GOSUB 230 ELSE GOSUB 260
160 NEXT J
170 U=H*(COS(F)+SIN(F))
180 V=H*(COS(F)-SIN(F))
190 GOSUB 210
200 NEXT N : NEXT M : A$=INPUT$(1) : END
210 LINE (X-V,Y-U)-(X+U,Y-V) : LINE -(X+V,Y+U)
220 LINE -(X-U,Y+V) : LINE -(X-V,Y-U) : RETURN
230 X=X-H*(COS(F)+2*SIN(F))
240 Y=Y+H*(2*COS(F)-SIN(F))
250 F=F+PI/4 : H=C*H : RETURN
260 X=X+H*(COS(F)-2*SIN(F))
270 Y=Y+H*(2*COS(F)+SIN(F))
280 F=F-PI/4 : H=C*H : RETURN
290 END



PYTHT2

10 REM ***LOPSIDED PYTHAGORAS TREE***
20 REM ***USING BINARY NUMBER SYSTEM***
30 REM ***NAME:PYTHT2***
40 SCREEN 3 : CLS : PI=3.141593
50 WINDOW (-2.5,-2)-(5.5,4)
60 DIM X(2048),Y(2048)
70 REM ***CHOICE OF ANGLE***
80 F=PI/3 : C=COS(F) : S=SIN(F)
90 A1=-C*S : A2=C^2 : B1=A1+A2 : B2=-A1+A2
100 C1=B2 : C2=1-B1 : D1=1-A1 : D2=1-A2
110 X(2)=0 : Y(2)=0 : X(3)=1 : Y(3)=0
120 LINE (0,0)-(1,0) : LINE -(1,-1) : LINE -(0,-1) :
    LINE -(0,0)
130 FOR M=1 TO 9
140 FOR J=0 TO 2^(M-1)-1
150 X0=X(2^M+2*J) : Y0=Y(2^M+2*J)
160 X1=X(2^M+2*J+1) : Y1=Y(2^M+2*J+1)
170 U=X1-X0 : V=Y1-Y0
180 XA=X0+A1*U-A2*V : YA=Y0+A2*U+A1*V
190 XB=X0+B1*U-B2*V : YB=Y0+B2*U+B1*V
200 XC=X0+C1*U-C2*V : YC=Y0+C2*U+C1*V
210 XD=X0+D1*U-D2*V : YD=Y0+D2*U+D1*V
220 X(2^(M+1)+4*J)=XA : Y(2^(M+1)+4*J)=YA
230 X(2^(M+1)+4*J+1)=XB : Y(2^(M+1)+4*J+1)=YB
240 X(2^(M+1)+4*J+2)=XC : Y(2^(M+1)+4*J+2)=YC
250 X(2^(M+1)+4*J+3)=XD : Y(2^(M+1)+4*J+3)=YD
260 LINE (X0,Y0)-(XA,YA) : LINE -(XB,YB)
270 LINE -(X1,Y1) : LINE -(XD,YD)
280 LINE -(XC,YC) : LINE -(X0,Y0)
```

```
290 NEXT J : NEXT M : BEEP
300 A$=INPUT$(1) : END
```

PYTHT3

```
10 REM ***PYTHAGORAS TREE, BACKTRACKING***
20 REM ***ORDER RESTRICTED BY DIAMETER OF SMALLEST SQUARE***
30 REM ***NAME:PYTHT3***
40 SCREEN 3 : CLS : PI=3.141593
50 WINDOW (-2.5,-2)-(5.5,4)
60 P=32 : DIM X2(P),Y2(P),U2(P),V2(P),S(P)
70 F=PI/3 : C=COS(F) : S=SIN(F) : REM***PRESCRIBED ANGLE***
80 EPS=.005 : REM ***TOLERANCE OF SMALLEST SQUARE***
90 A1=-C*S : A2=C^2 : B1=A1+A2 : B2=-A1+A2
100 C1=B2 : C2=1-B1 : D1=1-A1 : D2=1-A2
110 X1=0 : Y1=0 : U1=1 : V1=0 : Q=0 : J=1 : S(0)=1
120 LINE (0,0)-(1,0) : LINE -(1,-1) : LINE -(0,-1) :
    LINE -(0,0)
130 M=Q+J : X=U1-X1 : Y=V1-Y1
140 XA=X1+A1*X-A2*Y : YA=Y1+A2*X+A1*Y
150 XB=X1+B1*X-B2*Y : YB=Y1+B2*X+B1*Y
160 X2(M)=X1+C1*X-C2*Y : Y2(M)=Y1+C2*X+C1*Y
170 U2(M)=X1+D1*X-D2*Y : V2(M)=Y1+D2*X+D1*Y
180 S=X*X+Y*Y : S(M)=1
190 LINE (X1,Y1)-(XA,YA) : LINE -(XB,YB)
200 LINE -(U1,V1) : LINE -(U2(M),V2(M))
210 LINE -(X2(M),Y2(M)) : LINE -(X1,Y1)
220 X1=XA : Y1=YA : U1=XB : V1=YB
230 IF M=P OR S<EPS THEN GOSUB 250
240 J=J+1 : GOTO 130
250 K=1
260 IF S(M-K)=0 THEN K=K+1 : GOTO 260
270 IF M=K THEN A$=INPUT$(1) : END
280 Q=M-K : X1=X2(Q) : Y1=Y2(Q) : U1=U2(Q) : V1=V2(Q)
290 S(Q)=S(Q)-1 : J=0 : RETURN : END
```

PYTHTD

```
10 REM ***PYTHAGORAS TREE, BACKTRACKING***
20 REM ***WITH RANDOM DISTURBANCES***
30 REM ***NAME:PYTHTD***
40 SCREEN 3 : CLS : PI=3.141593 : RANDOMIZE 100
50 WINDOW (-9.5,-3)-(10.5,12) : W=0 : W=.15
60 P=10 : DIM X1(P),Y1(P),X2(P),Y2(P),U1(P),V1(P),U2(P),V2(P)
70 A1=0 : A2=3 : B1=.5 : B2=3.5 : C1=1 : C2=3
80 X1(0)=0 : Y1(0)=0 : U1(0)=1 : V1(0)=0
90 LINE (0,0)-(1,0)
100 S=1 : GOSUB 270 : GOSUB 170
110 FOR M=1 TO 2^(P-1)-1 : S=P : N=M
120 IF S<5 THEN GOSUB 270
130 IF N MOD 2=0 THEN N=N\2 : S=S-1 : GOTO 130
140 GOSUB 150 : NEXT M : BEEP : A$=INPUT$(1) : END
150 X1(S-1)=X2(S-1) : Y1(S-1)=Y2(S-1)
```

```
160 U1(S-1)=U2(S-1) : V1(S-1)=V2(S-1)
170 FOR J=S TO P
180 X=X1(J-1) : Y=Y1(J-1) : U=U1(J-1) : V=V1(J-1)
190 X3=U-X : Y3=V-Y
200 X1(J)=X+A1*X3-A2*Y3 : Y1(J)=Y+A2*X3+A1*Y3
210 U1(J)=X+B1*X3-B2*Y3 : V1(J)=Y+B2*X3+B1*Y3
220 X2(J)=U1(J) : Y2(J)=V1(J)
230 U2(J)=X+C1*X3-C2*Y3 : V2(J)=Y+C2*X3+C1*Y3
240 LINE (X,Y)-(X1(J),Y1(J)) :
    LINE (U1(J),V1(J))-(X2(J),Y2(J))
250 LINE (U2(J),V2(J))-(U,V)
260 NEXT J : RETURN : END
270 A2=A2*(1+(RND-.5)*W)
280 C2=C2*(1+(RND-.5)*W)
290 B2=(A2+C2)/2+.5
300 RETURN : END
```

SIER

```
10 REM ***SIERPINSKI SIEVE***
20 REM ***NAME:SIER***
30 SCREEN 3 : CLS : PI=3.141593
40 WINDOW (-2.6,-2.4)-(2.6,1.5)
50 P=5 : DIM T(P) : A=SQR(3)
60 FOR M=0 TO P
70 FOR N=0 TO 3^M-1
80 N1=N : FOR L=0 TO M-1
90 T(L)=N1 MOD 3 : N1=N1\3 : NEXT L
100 X=0 : Y=0
110 FOR K=0 TO M-1
120 X=X+COS((4*T(K)+1)*PI/6)/2^K
130 Y=Y+SIN((4*T(K)+1)*PI/6)/2^K
140 NEXT K
150 U1=X+A/2^(M+1) : U2=X-A/2^(M+1) : V1=Y-1/2^(M+1) :
    V2=Y+1/2^M
160 LINE (U1,V1)-(X,V2)
170 LINE -(U2,V1) : LINE -(U1,V1)
180 NEXT N : NEXT M : BEEP : A$=INPUT$(1) : END
```

SPHERSPI

```
10 REM ***SPHERICAL SPIRAL***
20 REM ***NAME:SPHERSPI***
30 SCREEN 3 : CLS : PI=3.141593
40 WINDOW (-2,-1.5)-(2,1.5)
50 A=.2 : REM ***SPIRAL PARAMETER***
60 C=.9 : REM ***SLOPE PROJECTION PLANE***
70 P=1/SQR(2) : Q=P*SQR(1-C*C) : REM ***PROJECTION
   CONSTANTS***
80 FOR N=-500 TO 500
90 S=N*PI/50 : T=ATN(A*S)
100 X=COS(S)*COS(T) : Y=SIN(S)*COS(T) : Z=-SIN(T)
110 U=P*(Y-X) : V=C*Z-Q*(X+Y)
```

```
120 IF N=-500 THEN  PSET(U,V) ELSE LINE -(U,V)
130 NEXT N : A$=INPUT$(1) : END
```

STAR

```
10 REM ***STAR FRACTAL***
20 REM ***NAME:STAR***
30 SCREEN 3 : CLS : PI=3.141593
40 WINDOW (-.5,-.8)-(1.5,.7)
50 P=5 : V=4 : A=144 : R=.35 : A=A*PI/180
60 PSET (0,0) : X=0 : Y=0
70 FOR N=0 TO (V+1)*V^(P-1)-1
80 M=N : B=N*A : F=0
90 IF M MOD V <> 0 OR F>=P-1 THEN GOTO 110
100 F=F+1 : M=INT(M/V) : GOTO 90
110 X=X+R^(P-F-1)*COS(B) : Y=Y+R^(P-F-1)*SIN(B)
120 LINE -(X,Y)
130 NEXT N : BEEP : A$=INPUT$(1) : END
```

TREE2

```
10 REM ***STRUCTURE OF A BINARY TREE***
20 REM ***NAME:TREE2***
30 SCREEN 3 : CLS
40 WINDOW (-2,-.5)-(2,2.5)
50 LINE (0,0)-(0,1)
60 FOR K=1 TO 7  : H=2^(-K+1)
70 FOR L=1 TO 2^K
80 X=-2+(4*L-2)*H : Y=2-H
90 LINE (X-H,Y+H/2)-(X-H,Y)
100 LINE (X-H,Y)-(X+H,Y) : LINE -(X+H,Y+H/2)
110 NEXT L : NEXT K
120 BEEP : A$=INPUT$(1) : END
```

TREE3

```
10 REM ***STRUCTURE OF A TERNARY TREE***
20 REM ***NAME:TREE3***
30 SCREEN 3 : CLS : PI=3.141593
40 WINDOW (-1.2,-.9)-(1.2,.9)
50 P=5 : DIM T(P) : REM ***ORDER***
60 A=.45 : REM ***REDUCTION***
70 FOR M=0 TO P
80 FOR N=0 TO 3^M-1
90 REM ***TERNARY NOTATION OF N***
100 N1=N : FOR L=1 TO M
110 T(L)=N1 MOD 3 : N1=N1\3 : NEXT L
120 X=0 : Y=0
130 FOR K=1 TO M
140 F=(2*T(K)*PI)/3
```

```
150 X=X+COS(F)*A^K : Y=Y+SIN(F)*A^K
160 NEXT K
170 LINE (X,Y)-(X+A^K,Y)
180 LINE (X,Y)-(X-.5*A^K,Y+SQR(3)/2*A^K)
190 LINE (X,Y)-(X-.5*A^K,Y-SQR(3)/2*A^K)
200 NEXT N : NEXT M
210 BEEP : A$=INPUT$(1) : END
```

TREEH1

```
10 REM ***H-FRACTAL***
20 REM ***NAME:TREEH1***
30 DIM X(2048),Y(2048)
40 SCREEN 3 : CLS
50 WINDOW (-2.4,-1.8)-(2.4,1.8)
60 P=9 : REM ***ORDER***
70 A=SQR(1/2) : REM ***REDUCTION***
80 X(1)=0 : Y(1)=0
90 FOR M=0 TO P : S=M MOD 2
100 FOR N=2^M TO 2^(M+1)-1
110 IF S=1 THEN GOSUB 160 ELSE GOSUB 190
120 NEXT N : NEXT M
130 FOR N=1 TO 2^(P+1)-1
140 LINE (X(2*N),Y(2*N))-(X(2*N+1),Y(2*N+1))
150 NEXT N : BEEP : A$=INPUT$(1) : END
160 X(2*N)=X(N) : Y(2*N)=Y(N)+A^M
170 X(2*N+1)=X(N) : Y(2*N+1)=Y(N)-A^M
180 RETURN
190 X(2*N)=X(N)+A^M : Y(2*N)=Y(N)
200 X(2*N+1)=X(N)-A^M : Y(2*N+1)=Y(N)
210 RETURN : END
```

TREEH2

```
10 REM ***H-FRACTAL, BACKTRACKING***
20 REM ***NAME:TREEH2***
30 SCREEN 3 : CLS
40 WINDOW (-1.2,-.9)-(1.2,.9)
50 P=6 : DIM X1(P),X2(P),X3(P),X4(P),Y1(P),Y2(P),Y3(P),Y4(P)
60 A=.5 : REM ***REDUCTION***
70 X1(0)=0 : Y1(0)=0 : S=1 : GOSUB 140
80 FOR M=1 TO 4^(P-1)-1 : N=M : S=P
90 IF N MOD 4 = 0 THEN N=N\4 : S=S-1 : GOTO 90
100 GOSUB 120 : NEXT M
110 BEEP : A$=INPUT$(1) : END
120 X1(S-1)=X2(S-1) : X2(S-1)=X3(S-1) : X3(S-1)=X4(S-1)
130 Y1(S-1)=Y2(S-1) : Y2(S-1)=Y3(S-1) : Y3(S-1)=Y4(S-1)
140 FOR J=S TO P
150 X=X1(J-1) : Y=Y1(J-1) : B=A^J : C=A*B*1.5
160 X1(J)=X+B : Y1(J)=Y+C
170 X2(J)=X+B : Y2(J)=Y-C
180 X3(J)=X-B : Y3(J)=Y+C
190 X4(J)=X-B : Y4(J)=Y-C
```

```
200 LINE (X-B,Y)-(X+B,Y)
210 LINE (X1(J),Y1(J))-(X2(J),Y2(J))
220 LINE (X3(J),Y3(J))-(X4(J),Y4(J))
230 NEXT J : RETURN : END
```

TREEM

```
10 REM ***MANDELBROT TREE, BACKTRACKING***
20 REM ***NAME:TREEM***
30 SCREEN 3 : CLS : PI=3.141593
40 WINDOW (-9.5,-3)-(10.5,12)
50 P=10 : DIM X1(P),Y1(P),X2(P),Y2(P),U1(P),V1(P),U2(P),V2(P)
60 R1=.72 : R2=.67 : A=3.98 : B=4.38
70 A1=0 : A2=A : B1=0 : B2=A+R1
80 E1=1 : E2=B+R2 : F1=1 : F2=B
90 C1=.5 : C2=B2 : D1=.5 : D2=E2
100 X1(0)=0 : Y1(0)=0 : U1(0)=1 : V1(0)=0
110 LINE (0,0)-(1,0)
120 S=1 : GOSUB 210
130 FOR M=1 TO 2^(P-1)-1 : S=P : N=M
140 IF N MOD 2=0 THEN N=N\2 : S=S-1 : GOTO 140
150 H=A2 : A2=F2 : F2=H : H=B2 : B2=E2 : E2=H :
    H=C2 : C2=D2 : D2=H
160 GOSUB 190
170 NEXT M : BEEP
180 A$=INPUT$(1) : END
190 X1(S-1)=X2(S-1) : Y1(S-1)=Y2(S-1)
200 U1(S-1)=U2(S-1) : V1(S-1)=V2(S-1)
210 FOR J=S TO P
220 X=X1(J-1) : Y=Y1(J-1) : U=U1(J-1) : V=V1(J-1)
230 X3=U-X : Y3=V-Y
240 X1(J)=X+A1*X3-A2*Y3 : Y1(J)=Y+A2*X3-A1*Y3
250 U1(J)=X+B1*X3-B2*Y3 : V1(J)=Y+B2*X3+B1*Y3
260 X2(J)=X+E1*X3-E2*Y3 : Y2(J)=Y+E2*X3+E1*Y3
270 U2(J)=X+F1*X3-F2*Y3 : V2(J)=Y+F2*X3+F1*Y3
280 U3=X+C1*X3-C2*Y3 : V3=Y+C2*X3+C1*Y3
290 U4=X+D1*X3-D2*Y3 : V4=Y+D2*X3+D1*Y3
300 IF J=S THEN H=A2 : A2=F2 : F2=H : H=B2 : B2=E2 : E2=H :
    H=C2 : C2=D2 : D2=H
310 LINE (X,Y)-(X1(J),Y1(J)) : LINE (U1(J),V1(J))-(U3,V3)
320 LINE -(U4,V4) : LINE -(X2(J),Y2(J)) :
    LINE (U2(J),V2(J))-(U,V)
330 NEXT J : RETURN : END
```

UNWIND

```
10 REM ***EVOLUTE CIRCLE***
20 REM ***NAME:UNWIND***
30 SCREEN 3 : CLS : PI=3.141593
40 WINDOW (-12,-9)-(12,9)
50 A=1.2 : REM ***RADIUS OF CIRCLE***
60 CIRCLE (0,0),A : PSET (A,0)
70 FOR N=0 TO 100 : T=2*PI*N/80
```

```
 80 X=A*(COS(T)+T*SIN(T))
 90 Y=A*(SIN(T)-T*COS(T))
100 LINE -(X,Y)
110 IF N MOD 10 = 0 THEN LINE (A*COS(T),A*SIN(T))-(X,Y)
120 NEXT N : PAINT (0,0)
130 A$=INPUT$(1) : END
```

WHIRL

```
 10 REM ***ROTATING AND REDUCING POLYGON***
 20 REM ***NAME:WHIRL***
 30 SCREEN 3 : CLS : PI=3.141593
 40 WINDOW (-4/3,-1)-(4/3,1)
 50 PRINT "SELECT NUMBER OF SIDES"
 60 INPUT P : CLS : DIM X(P),Y(P)
 70 B=.05 : REM ***ROTATION ANGLE IN RADIALS***
 80 A=PI*(1-2/P) : C=SIN(A)/(SIN(B)+SIN(A+B))
 90 FOR K=0 TO P : T=(2*K+1)*PI/P
100 X(K)=SIN(T) : Y(K)=COS(T)
110 NEXT K
120 FOR N=1 TO 64  : PSET (X(0),Y(0))
130 FOR L=1 TO P : LINE -(X(L),Y(L))
140 NEXT L
150 FOR M=0 TO P : Z=X(M)
160 X(M)=(X(M)*COS(B)-Y(M)*SIN(B))*C
170 Y(M)=(Z   *SIN(B)+Y(M)*COS(B))*C
180 NEXT M : NEXT N
190 BEEP : A$=INPUT$(1) : END
```

BIBLIOGRAPHY

Lauwerier, H. A. *Analyse met de microcomputer*. Epsilon uitgaven, 1987.

Lauwerier, H. A. *Meetkunde met de microcomputer*. Epsilon uitgaven, 1987.
These volumes are aimed at use in schools. The emphasis is on making graphic representations, tracing figures in the plane, and projections of spatial bodies. They include a total of ninety computer programs. The second column ends with a chapter on the fourth dimension.

Lauwerier, H. A. *Symmetrie: Regelmatige structuren in de kunst*. Aramith uitgevers, 1988.

Mandelbrot, B. B. *The Fractal Geometry of Nature*. Freeman, 1983.
An essay with the emphasis on the principle of repetition. It contains many illustrations.

Peitgen, H.-O., and P. H. Richter. *The Beauty of Fractals*. Springer, 1986.
Attractive because of the great number of colored illustrations. Emphasis on the fractals of Julia and Mandelbrot. Includes tough mathematics.

Schuster, H. G. *Deterministic Chaos: an Introduction*. Physik-Verlag, 1984.
General introduction to chance in fractals, dynamical systems, and the fractals of Julia and Mandelbrot. For the scientifically minded reader. It contains mathematics, though it lacks depth. Links up with Chapters 6 and 7 of our book.

INDEX